QUALITY ASSURANCE AND OPTIMIZATION FOR FLUOROSCOPICALLY GUIDED INTERVENTIONAL PROCEDURES

The following States are Members of the International Atomic Energy Agency:

AFGHANISTAN
ALBANIA
ALGERIA
ANGOLA
ANTIGUA AND BARBUDA
ARGENTINA
ARMENIA
AUSTRALIA
AUSTRIA
AZERBAIJAN
BAHAMAS
BAHRAIN
BANGLADESH
BARBADOS
BELARUS
BELGIUM
BELIZE
BENIN
BOLIVIA, PLURINATIONAL STATE OF
BOSNIA AND HERZEGOVINA
BOTSWANA
BRAZIL
BRUNEI DARUSSALAM
BULGARIA
BURKINA FASO
BURUNDI
CABO VERDE
CAMBODIA
CAMEROON
CANADA
CENTRAL AFRICAN REPUBLIC
CHAD
CHILE
CHINA
COLOMBIA
COMOROS
CONGO
COOK ISLANDS
COSTA RICA
CÔTE D'IVOIRE
CROATIA
CUBA
CYPRUS
CZECH REPUBLIC
DEMOCRATIC REPUBLIC OF THE CONGO
DENMARK
DJIBOUTI
DOMINICA
DOMINICAN REPUBLIC
ECUADOR
EGYPT
EL SALVADOR
ERITREA
ESTONIA
ESWATINI
ETHIOPIA
FIJI
FINLAND
FRANCE
GABON
GAMBIA, THE
GEORGIA
GERMANY
GHANA
GREECE
GRENADA
GUATEMALA
GUINEA
GUYANA
HAITI
HOLY SEE
HONDURAS
HUNGARY
ICELAND
INDIA
INDONESIA
IRAN, ISLAMIC REPUBLIC OF
IRAQ
IRELAND
ISRAEL
ITALY
JAMAICA
JAPAN
JORDAN
KAZAKHSTAN
KENYA
KOREA, REPUBLIC OF
KUWAIT
KYRGYZSTAN
LAO PEOPLE'S DEMOCRATIC REPUBLIC
LATVIA
LEBANON
LESOTHO
LIBERIA
LIBYA
LIECHTENSTEIN
LITHUANIA
LUXEMBOURG
MADAGASCAR
MALAWI
MALAYSIA
MALI
MALTA
MARSHALL ISLANDS
MAURITANIA
MAURITIUS
MEXICO
MONACO
MONGOLIA
MONTENEGRO
MOROCCO
MOZAMBIQUE
MYANMAR
NAMIBIA
NEPAL
NETHERLANDS, KINGDOM OF THE
NEW ZEALAND
NICARAGUA
NIGER
NIGERIA
NORTH MACEDONIA
NORWAY
OMAN
PAKISTAN
PALAU
PANAMA
PAPUA NEW GUINEA
PARAGUAY
PERU
PHILIPPINES
POLAND
PORTUGAL
QATAR
REPUBLIC OF MOLDOVA
ROMANIA
RUSSIAN FEDERATION
RWANDA
SAINT KITTS AND NEVIS
SAINT LUCIA
SAINT VINCENT AND THE GRENADINES
SAMOA
SAN MARINO
SAUDI ARABIA
SENEGAL
SERBIA
SEYCHELLES
SIERRA LEONE
SINGAPORE
SLOVAKIA
SLOVENIA
SOMALIA
SOUTH AFRICA
SPAIN
SRI LANKA
SUDAN
SWEDEN
SWITZERLAND
SYRIAN ARAB REPUBLIC
TAJIKISTAN
THAILAND
TOGO
TONGA
TRINIDAD AND TOBAGO
TUNISIA
TÜRKİYE
TURKMENISTAN
UGANDA
UKRAINE
UNITED ARAB EMIRATES
UNITED KINGDOM OF GREAT BRITAIN AND NORTHERN IRELAND
UNITED REPUBLIC OF TANZANIA
UNITED STATES OF AMERICA
URUGUAY
UZBEKISTAN
VANUATU
VENEZUELA, BOLIVARIAN REPUBLIC OF
VIET NAM
YEMEN
ZAMBIA
ZIMBABWE

The Agency's Statute was approved on 23 October 1956 by the Conference on the Statute of the IAEA held at United Nations Headquarters, New York; it entered into force on 29 July 1957. The Headquarters of the Agency are situated in Vienna. Its principal objective is "to accelerate and enlarge the contribution of atomic energy to peace, health and prosperity throughout the world".

IAEA HUMAN HEALTH SERIES No. 48

QUALITY ASSURANCE AND OPTIMIZATION FOR FLUOROSCOPICALLY GUIDED INTERVENTIONAL PROCEDURES

ENDORSED BY THE AMERICAN ASSOCIATION OF PHYSICISTS IN MEDICINE, ASIA-OCEANIA FEDERATION OF ORGANIZATIONS FOR MEDICAL PHYSICS, ASOCIACIÓN LATINOAMERICANA DE FÍSICA MÉDICA, EUROPEAN FEDERATION OF ORGANISATIONS FOR MEDICAL PHYSICS, FEDERATION OF AFRICAN MEDICAL PHYSICS ORGANIZATIONS, INTERNATIONAL ORGANIZATION FOR MEDICAL PHYSICS, INTERNATIONAL SOCIETY OF RADIOGRAPHERS AND RADIOLOGICAL TECHNOLOGISTS, AND SOUTHEAST ASIA FEDERATION OF ORGANIZATIONS FOR MEDICAL PHYSICS

INTERNATIONAL ATOMIC ENERGY AGENCY
VIENNA, 2025

COPYRIGHT NOTICE

Publishing Section
International Atomic Energy Agency
Vienna International Centre
PO Box 100
1400 Vienna, Austria
tel.: +43 1 2600 22529 or 22530
email: sales.publications@iaea.org
www.iaea.org/publications

Printed by the IAEA in Austria
March 2025
STI/PUB/2101
https://doi.org/10.61092/iaea.f7bd-coh3

IAEA Library Cataloguing in Publication Data

Names: International Atomic Energy Agency.
Title: Quality assurance and optimization for fluoroscopically guided interventional procedures / International Atomic Energy Agency.
Description: Vienna : International Atomic Energy Agency, 2025. | Series: IAEA human health series, ISSN 2075–3772 ; no. 48 | Includes bibliographical references.
Identifiers: IAEAL 24-01723 | ISBN 978-92-0-134524-0 (paperback : alk. paper) | ISBN 978-92-0-134724-4 (pdf) | ISBN 978-92-0-134624-7 (epub)
Subjects: LCSH: Fluoroscopy. | Fluoroscopy — Equipment and supplies. | Radiation — Safety measures. | Diagnostic imaging. | Quality assurance.
Classification: UDC 614.876 | STI/PUB/2101

FOREWORD

Fluoroscopically guided interventional procedures offer a minimally invasive approach to diagnosing and treating various clinical conditions. These procedures are performed using intravascular percutaneous or endoscopic techniques, providing a safer, less invasive and often more cost effective alternative to open surgery. They are applicable to both adult and paediatric patients. Clinical departments that utilize fluoroscopically guided interventional procedures include interventional radiology, interventional cardiology, cardiac electrophysiology, interventional neuroradiology, gastroenterology, musculoskeletal radiology, gynaecology, pulmonology, urology, oncologic surgery, orthopaedics and vascular surgery.

The deployment of advanced X ray fluoroscopic imaging equipment is indispensable for implementing these procedures. These systems are equipped with multiple advanced clinical protocols and imaging tools that are tailored to accommodate procedure specific imaging needs. Additionally, they offer the flexibility of higher radiation dose and higher image quality, where required, as well as enhanced functionality for more precise radiation dose tracking. Proper training in the safe and effective use of this cutting edge equipment is also paramount for clinical specialists, ensuring that these capabilities are properly utilized to provide optimal patient care.

As the application and complexity of fluoroscopically guided interventional procedures continue to grow, the collaboration between various clinical professionals is of increased importance. Together, these clinical professionals ideally will not only establish a quality assurance programme but also develop comprehensive optimization processes tailored to the unique demands of fluoroscopically guided interventional procedures. The quality assurance programme sets the groundwork for consistent, high quality performance across X ray fluoroscopic imaging systems by defining standards, procedures and guidelines that align with best practices and regulatory requirements. It focuses on the monitoring, evaluation and improvement of all aspects related to fluoroscopically guided interventional procedures, from equipment maintenance to protocol adherence.

Optimization processes, on the other hand, involve a systematic approach to enhancing the efficiency, effectiveness and safety of fluoroscopic procedures. By fine tuning settings, customizing clinical protocols and leveraging the latest technological advancements, these processes work to balance the need for good quality imaging with the imperative to minimize radiation exposure to both patients and medical staff. Through these collaborative efforts, clinical professionals forge a partnership that elevates patient care, ensures adherence to regulatory standards and fosters continuous innovation in the rapidly evolving field of fluoroscopically guided interventional procedures. Their combined expertise ensures that X ray fluoroscopic imaging systems are utilized to their full potential, delivering precise diagnoses and interventions while prioritizing patient safety.

The IAEA has already issued guidance on quality control, quality assurance and dosimetry for various radiological modalities, although not specifically for fluoroscopically guided interventional procedures. To address the need for guidance in this area, the present publication was developed taking into account the recommendations of the Scientific Committee and represents the current state of quality control, quality assurance as well as optimization of X ray equipment and procedures.

The IAEA acknowledges the contribution of the drafting committee responsible for the development of this publication as well as the review of internal and external experts listed at the end of the publication. The committee includes H.J. de Vos (South Africa), C. Maccia (France), F. Malchair (Belgium), Chaihong Yeong (Malaysia), K.A. Wunderle (United States of America) and C. Ubeda (Chile). The IAEA officer responsible for this publication was V. Tsapaki of the Division of Human Health.

This publication is endorsed by the American Association of Physicists in Medicine, Asia-Oceania Federation of Organizations for Medical Physics, Asociación Latinoamericana de Física Médica, European Federation of Organisations for Medical Physics, Federation of African Medical Physics Organizations, International Organization for Medical Physics, International Society of Radiographers and Radiological Technologists, and Southeast Asia Federation of Organizations for Medical Physics.

CONTENTS

1. INTRODUCTION

1.1. BACKGROUND

The value of fluoroscopically guided interventional procedures (FGIPs) in medical practice is undeniable [1, 2]. These procedures offer an efficient means to diagnose and treat various clinical problems, typically resulting in reduced complications and mortality rates, expedited diagnosis and treatment, shorter hospital stays and faster overall recovery. They often present a more cost effective alternative compared to other surgical or medical interventions. Not only do they tend to be less expensive, but they also usually lead to reduced complications and morbidity for patients. Furthermore, FGIPs often result in shorter hospital stays, minimizing the overall impact on patients' lives and potentially leading to quicker recoveries. This combination of lower costs, decreased patient discomfort and reduced time spent in hospital settings makes FGIPs a preferred choice in many medical scenarios where they are applicable.

As global populations expand, there is a corresponding increase in the number of congenital and chronic conditions that can be effectively addressed by FGIPs. This trend emphasizes the growing need for both the availability of these advanced services and robust quality assurance (QA)/quality control (QC) programmes. These programmes ensure that FGIPs are applied appropriately and efficaciously, especially as they become more essential in treating an ever-increasing number of conditions. The mounting demand for FGIPs underscores their vital role in modern healthcare, offering efficient and less invasive solutions to a wider array of patients. Simultaneously, it highlights the importance of maintaining rigorous quality standards, reinforcing the necessity for ongoing vigilance in QA and patient safety.

Over the past few decades, significant advancements in medical technology have revolutionized the way patient data is managed and utilized. Information that once required physical storage devices such as compact disc read-only memory can now be accessed effortlessly through server-based technology. This technological shift has enabled quality assurance programmes (QAPs) to not only track radiation dose data with greater ease but also participate in clinical or technical registries. This allows for the comparison of local data with larger datasets, fostering an environment of continuous learning, improvement and high reliability.

The complexity of FGIPs, however, can bring increased risks, such as a higher patient radiation dose that may lead to tissue reactions. This underscores the necessity for stringent QC/QA, precise optimization processes and unwavering commitment to minimizing radiation exposure. All related collaboration between clinical professionals and radiological medical practitioners is crucial in this regard. Together, they ensure that FGIPs are performed with the highest standards of safety and efficacy, aligning technology and practice with patient wellbeing.

The IAEA recognizes the importance of this comprehensive approach to quality in diagnostic imaging. The organization actively promotes the development of guidance and training for the appropriate staff involved in maintaining this structure of quality and safety. Through these collaborative efforts, the medical community ensures that FGIPs are conducted with the highest standards of efficiency, effectiveness and patient safety.

1.2. OBJECTIVE

The aim of the present publication is to provide comprehensive guidance on QA and on optimization in FGIPs. This guidance is intended to establish a framework that ensures the highest standards of safety, effectiveness, and efficiency in FGIPs. By outlining best practices, standardizing protocols and emphasizing the integration of cutting edge technology with evidence based medicine, the publication aims to support healthcare providers in offering FGIPs that are patient-centred and in line with global standards. The focus on QA and optimization encompasses not only the technical aspects of FGIPs but also extends to training, collaboration among medical professionals, data management, and continuous evaluation and improvement. Ultimately, this guidance seeks to foster an environment where QC is integral to the delivery of FGIPs, enhancing patient outcomes and advancing the field of interventional medicine. Guidance and recommendations provided here in relation to identified good practices represent expert opinion but are not made on the basis of a consensus of all Member States.

1.3. SCOPE

This publication specifically centres on the development and implementation of QAP in FGIPs as well as comprehensive optimization process. Guidelines are provided for conducting QC tests, with attention to common issues and mistakes that could compromise results. Detailed instructions for performing each test are included to ensure accuracy and reliability. The publication also emphasizes the standards for evaluating clinical images, offering guidance on the criteria and methodologies to assess the adequacy of the images in a clinical context. Finally, the factors influencing image quality and radiation exposure, including actionable guidance and clinical examples for daily practice are outlined. The aim is to strike a balance between image clarity and minimizing patient exposure to radiation. Moreover, fluoroscopy and another image modality or modalities may be utilized during a singular procedure in a complementary, synergistic fashion, which is often the case for ultrasound and fluoroscopy, as well as for the placement of central venous catheters.

It is important to note that the scope of this publication is restricted to fluoroscopy. It does not encompass any aspects related to the use of radioactive materials, even for procedures performed in an FGIP suite that may involve substances such as Y-90 radioembolizations. The complexities tied to those procedures — including additional knowledge, training, policies and regulatory oversight — are beyond the scope of this publication. This focus ensures that the guidance provided remains pertinent and applicable to the specific context of FGIPs, enhancing the overall quality and safety of these essential medical procedures.

1.4. STRUCTURE

This publication is divided into seven sections, each dealing with a different topic. Section 1 is an introduction to the subject. Section 2 outlines main clinical applications of FGIPs. Section 3 provides the latest updates on fluoroscopy and angiography technology. Section 4 includes the basic principles of QAP in FGIPs. Section 5 provides the relevant performance tests. Section 6 gives an overview of optimization process including both image quality and radiation exposure, and Section 7 outlines the conclusions of the guidelines. The last part of the publication contains four (4) appendices that provide a list of FGIPs (Appendix I), uncertainties in measurements (Appendix II), information on patient dose assessment process (Appendix III), and examples of optimization processes in the everyday clinical practice for cardiology (Appendix IV-A), cardiac electrophysiology (Appendix IV-B) and endoscopic retrograde cholangiopancreatography (ERCP) (Appendix IV-C).

2. CLINICAL APPLICATIONS

This section delves into the various clinical applications where FGIPs play a pivotal role. These procedures span across numerous clinical specialties, highlighting the versatility and importance of this imaging modality. Below are some common procedures in different clinical areas where fluoroscopy provides indispensable assistance, and Appendix I provides more examples of the vast array of procedures that may employ fluoroscopic guidance.

2.1. CARDIAC

Several clinical applications of FGIPs exist in the general field of cardiology, each targeting specific issues such as haemodynamic disturbances, electrical malfunctions or valvular complications, whether congenital or contracted. These innovative applications have contributed to a reduction in the need for open heart surgeries, enabling radiological medical practitioners to treat patients through less invasive methods. These approaches often yield comparable or even improved outcomes, reflecting substantial advancements in cardiac care.

Modern cardiac procedures include:

(a) Haemodynamic procedures: Coronary angiography, percutaneous coronary interventions including treatment of chronic, acute or subacute coronary occlusion of varying degrees, as shown in Fig. 1.
(b) Electrophysiology procedures: Electrical device implantation (such as pacemaker, defibrillator, Holter, etc.), tissue ablations for aberrant rhythms or electrical signal malfunctions.
(c) Structural procedures: Patent foramen oval closure, cardiac valve (e.g. aortic, mitral, tricuspid, bicuspid) replacement or repair, as shown in Fig. 2.

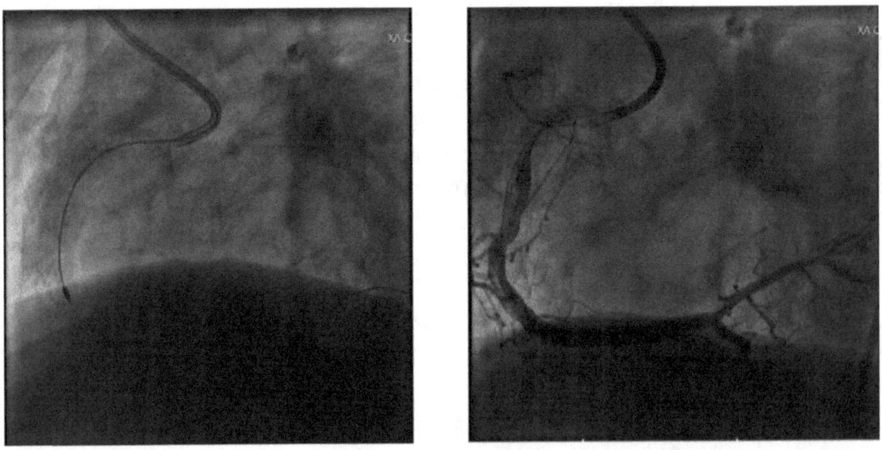

(a) (b)

FIG. 1. Chronic total occlusion of right coronary artery: (a) before angioplasty and (b) post angioplasty.

(d) Specific paediatric procedures: congenital heart defects (e.g. patent foramen ovale, tetralogy of Fallot, atrial or ventricular septal defects, etc.).

2.2. NEUROVASCULAR

FGIP techniques are frequently used in the diagnosis and treatment of various neurovascular pathologies that affect the blood vessels in the brain, head, neck and spine. The X ray guided endovascular approach is less invasive than conventional surgery, which reduces patient hospitalisation and recovery times considerably. Neurovascular procedures may be done by interventional radiologists, neurointerventional radiologists or neurosurgeons [3].

Some common conditions treated using neurointerventional radiology include arteriovenous malformations brain aneurysms, brain tumour embolization, carotid artery disease and stroke. Figure 3 shows an example of an arteriovenous malformation case, and Fig. 4 an embolization case. In particular, this non-invasive endovascular approach makes FGIPs popular for the treatment of various paediatric neurovascular pathologies like complex haemangiomas, vascular abnormalities and neurovascular diseases found in children.

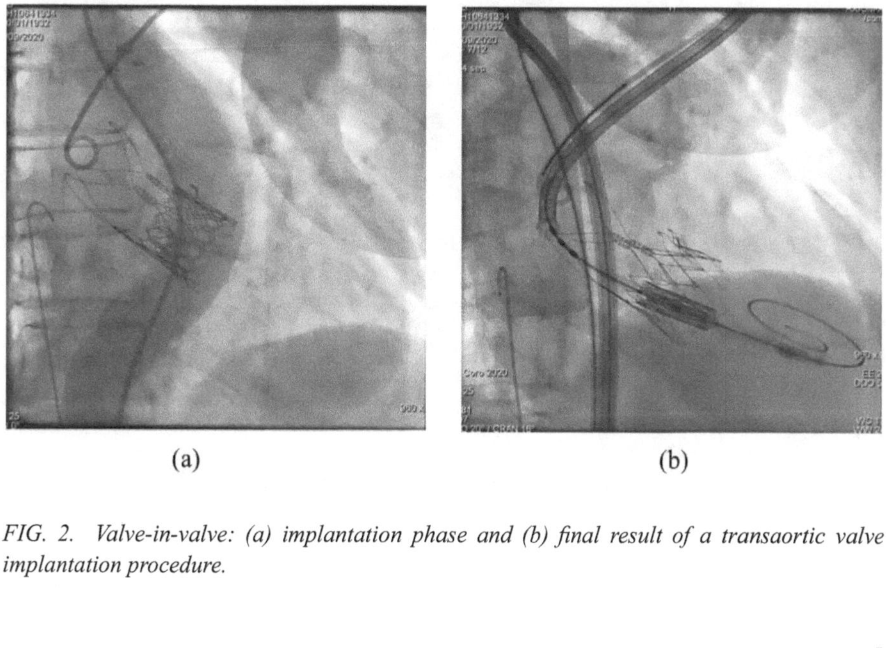

(a) (b)

FIG. 2. Valve-in-valve: (a) implantation phase and (b) final result of a transaortic valve implantation procedure.

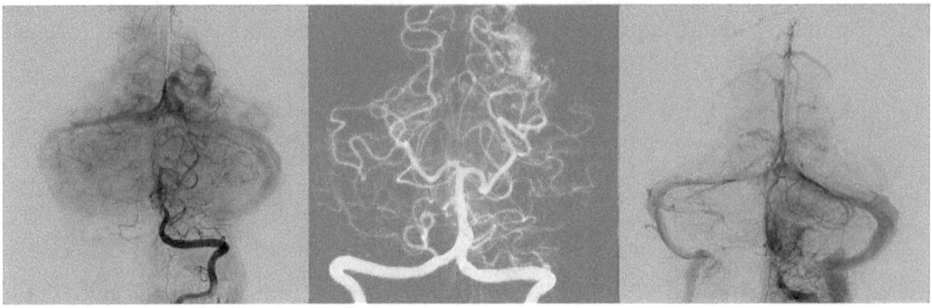

FIG. 3. Examples of arteriovenous malformation cases.

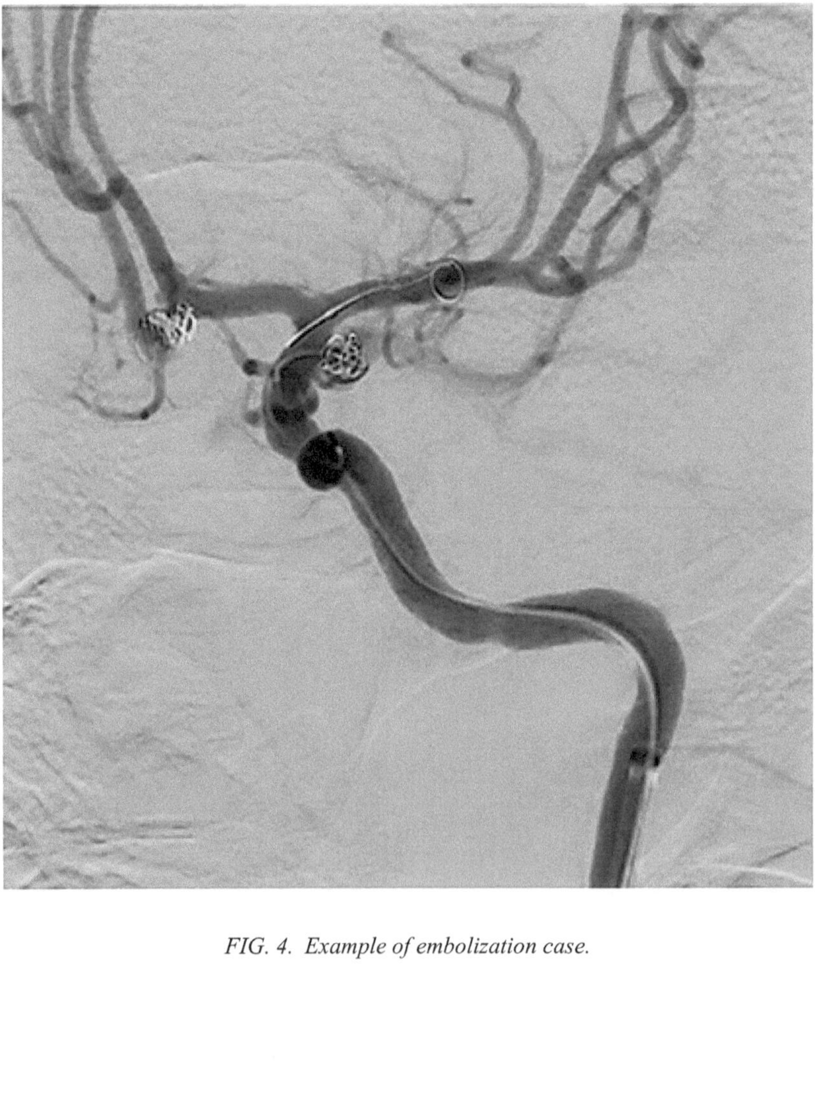

FIG. 4. Example of embolization case.

2.3. PERIPHERAL VASCULAR

Peripheral vascular diseases are widespread conditions that can affect extensive regions within the human body. Historically, treating these diseases required invasive open surgical methods that were characterized by large incisions. These traditional approaches often led to painful recoveries and prolonged healing periods. In contrast, peripheral vascular FGIPs provide a less invasive option. These techniques utilize small surgical incisions, allowing for catheter access to the targeted area needing treatment. With the aid of X ray fluoroscopy equipment, medical professionals can precisely position catheters and other instruments such as needles, balloon catheters, prostheses and various devices. Figure 5 illustrates an example of a peripheral vascular FGIP case. These procedures often go beyond the use of X ray fluoroscopy alone. They may incorporate additional imaging guidance systems or a synergistic combination of different imaging and navigation technologies. This could include ultrasound, magnetic guidance systems or computed tomography (CT) systems. The integration of these advanced tools enhances the precision and efficacy of interventions, making peripheral vascular FGIPs a valuable advancement in the treatment of vascular diseases, minimizing invasiveness and reducing recovery times [3].

One of these FGIP procedures is endovascular aneurysm repair. Endovascular aneurysm repair represents a minimally invasive method characterized by the internal relining of the aorta using a stent graft, thereby excluding the aneurysm from systemic circulation. This procedure necessitates precise X ray guided placement and manipulation of the stent graft within the aortic structure. With advancements in this technique, there is a notable increase in the utilization and complexity of endovascular therapies for the treatment of aortic often involving the deployment of multiple stent components in extended procedures that may last for several hours. Given the procedural intricacies and the inherent radiodensity of the abdominal region, endovascular aneurysm repair is predisposed to the administration of significantly high radiation doses.

2.4. INTERVENTIONAL ONCOLOGY

Interventional oncology has rapidly evolved over the last several decades, emerging as a vital branch of FGIPs. This discipline specifically targets solid cancers, utilizing localized approaches to attack tumours, rather than using systemic treatments that affect the entire body. Interventional oncology methods are primarily divided into two categories: transvascular and percutaneous approaches.

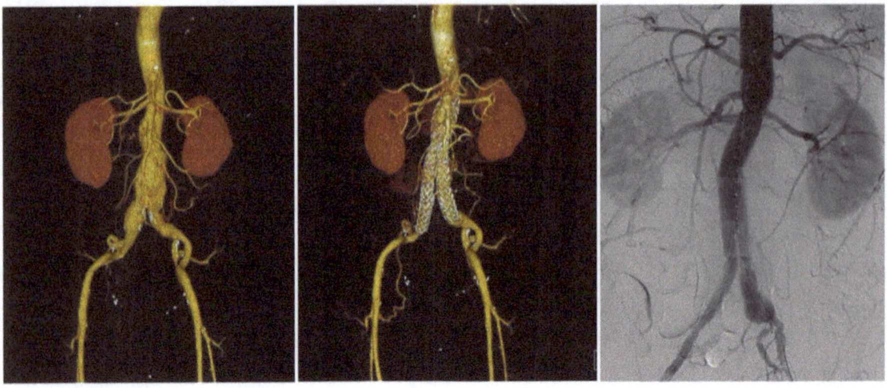

FIG. 5. Pre- and post-stent graft in aorta.

(a) Transvascular approaches: These methods leverage the vascular structures within the body and tumour for precise access. The strategies often begin with embolization, which blocks the blood supply to the tumour (bland embolization). Additionally, there may be direct administration of chemotherapeutic drugs to the tumour site (chemoembolization) or the administration of radioactive embolic agents for internal irradiation (radioembolization) in addition to embolization.

(b) Percutaneous approaches: These techniques typically employ various forms of ablative procedures, such as radiofrequency, microwave, cryoablation or focused ultrasound. These methods thermally or mechanically damage the targeted tissues, providing another route to attack the cancer.

The field also benefits from continuous advancements in targeted immunotherapy compounds, which further enriches the therapeutic options available. These innovative treatments are being introduced and refined at an accelerated pace, ensuring that interventional oncology continues to be at the forefront of cancer treatment. In the setting of advanced cervical cancer encasing the ureters, both ultrasound and fluoroscopy are used synergistically during percutaneous nephrostomy and ureteral stent procedures. Basic interventional radiology also includes the placement of central venous catheters (ports or peripherally inserted central catheters (PICC lines)), which are requisite for the administration of many systemic agents used in cancer therapeutic regimens, including several WHO Essential Medicines that are included among anti-neoplastic agents. Like many interventional procedures, this catheter placement generally uses both fluoroscopy and ultrasound synergistically.

Overall, interventional oncology represents a significant advancement in cancer care, offering targeted and localized treatments, several examples of which have been provided above. Its growth is projected to continue accelerating, reflecting the dynamic and essential nature of this clinical discipline [4]. Figure 6 shows an example of an interventional oncology procedure.

2.5. GASTROENTEROLOGY

Basic FGIPs in gastroenterology include upper GI studies, with oral contrast, as well small bowel follow-throughs and contrast-enhanced colonic studies undertaken with enemas. In addition, FGIPs may be used for fistulograms or to assess post-operative anastomotic leaks. ERCP is the most conducted procedure in gastroenterology that utilizes ionizing radiation. Other procedures in this domain involve digestive tract stent insertion, strictures dilation, small bowel enteroscopy, transluminal interventional endoscopic ultrasonography and insertion of enteric feeding tubes, among others. Figure 7 presents a sample ERCP case. The execution of FGIPs in gastroenterology is commonly carried out by medical professionals across various specialties, including gastroenterologists, surgeons and interventional radiologists. These procedures can be performed in a diverse range of settings such as within the radiology department utilizing either an over-the-couch or under-the-couch X ray tube system, in an operating theatre, in the endoscopy unit with a mobile C-arm or in an interventional imaging suite employing an angiography system.

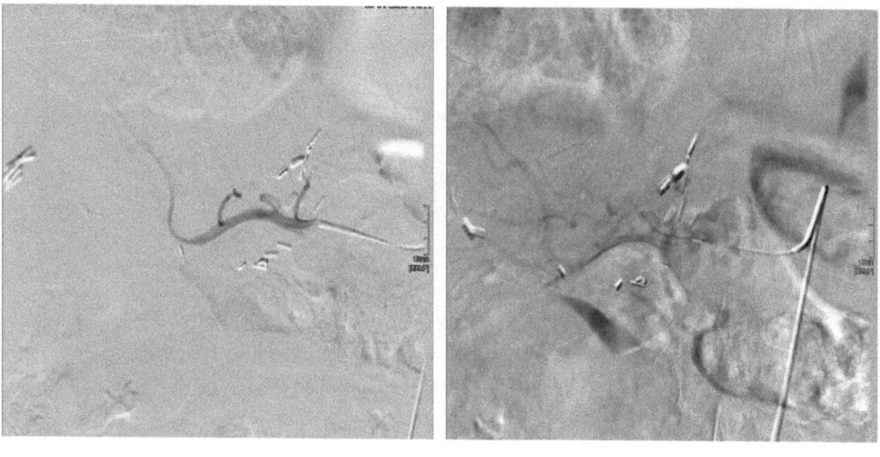

FIG. 6. Example of an interventional oncology procedure (chemoembolization).

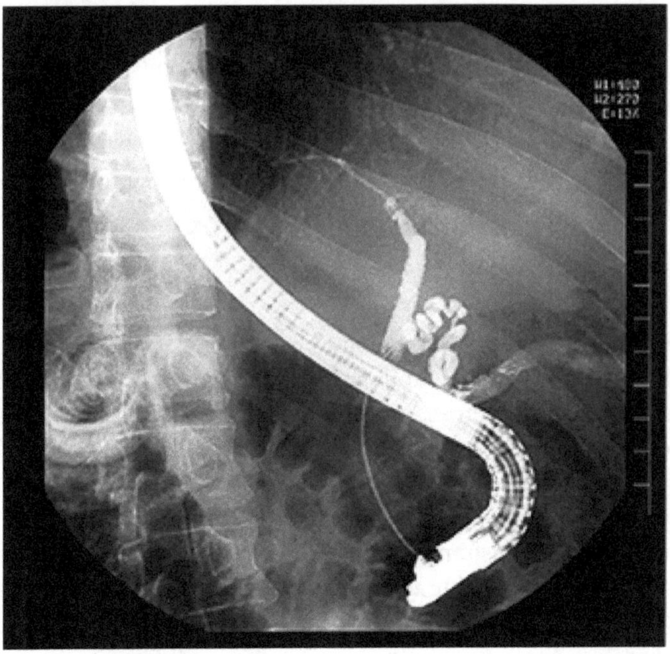

FIG. 7. An example of ERCP case.

3. FGIP TECHNOLOGY

This section offers a comprehensive exploration of the key factors critical to the design and functionality of an FGIP facility. It covers a spectrum of topics, ranging from the diverse architectural considerations and technological configurations of FGIP imaging systems to their intricate components, as well as a thorough consideration of additional factors that influence operational efficiency, image quality and patient safety.

3.1. BASIC CONSIDERATIONS

The planning and design of a new FGIP facility and related dependencies are complex and expensive processes. At a minimum, the team ideally should comprise a senior physician trained in these procedures, also called a radiological medical practitioner, a clinically qualified medical physicist (CQMP), a senior medical radiation technologist (MRT), also known as a radiographer, and a facility engineer [5]. In the case that the facility also has a radiation protection officer in addition to the CQMP, then he or she ideally needs to be included, as well.

Facility design is an essential component of radiation safety and management. Examples of architectural suggestions for cardiac catheterisation laboratories have been published by other scientific organizations and authors [6, 7]. As stated in National Council on Radiation Protection and Measurements (NCRP) Report No. 168 [2], having an ample sized procedure room is advantageous, as it not only provides ancillary staff with sufficient space to manoeuvre and operate safely during procedures but also facilitates easy access for patients on trolley beds. Additionally, a spacious room allows for easier patient positioning and enables more fluid movement of both equipment and patients. This enhanced mobility not only helps in reducing the time and medical exposure during the procedure but also aids in minimizing scatter and leakage radiation, thereby reducing occupational exposure. A large control room with a generously sized shielded glass window can serve as an observation area. This permits non-essential staff to remain in a shielded area. An intercom between the control and procedure rooms can also allow some staff to remain outside of the procedure room until they are needed. Designs that permit easy and comfortable access to mobile and fixed shielding increase the likelihood that this shielding will be used.

As specified by the second edition of International Electrotechnical Commission (IEC) Standard 60601-2-54 [8], an audible alert needs to be triggered after 5 min of fluoroscopy time has elapsed or any time the system is operated in high dose fluoroscopy mode. It is also important that lights or signs be available at the entrance and outside each entry into a procedure room to indicate that the X ray equipment is in use or X ray beam is activated so that precautions ought to be taken if room entry is required [2]. In addition, it is advisable to use caution lights of universal design (e.g. amber light) at multiple locations inside the procedure room so that they are visible to all staff working anywhere in the room. These lights are continuously illuminated while X rays are being produced [2].

3.2. DESCRIPTION OF THE FGIP SYSTEMS

This guidance is intended to address fluoroscopic systems meeting the IEC 60601-2-43 standard entitled “Particular requirements for the basic safety and essential performance of X-ray equipment for interventional procedures” [9]. Fluoroscopic systems meeting the IEC requirements of this standard are most often fixed or stationary C-arm type fluoroscopic systems, although some mobile C-arm systems also meet this standard. Only a few of the IEC requirements related to FGIP equipment are listed below, but it is important for CQMP to know and understand these features, and to train clinical colleagues on the need for their routine use.

The following is a non-exhaustive list of requirements for the FGIP equipment on the basis of the IEC standard requirements:

(a) Produce and be capable of transmitting a radiation dose structured report (RDSR). These reports contain both summary radiation dose indices (e.g. air kerma at reference point [$K_{a,r}$] and air kerma-area product [P_{KA}]) as well as irradiation event level data for each radiation exposure during an FGIP procedure. This event level data includes fluoroscopy equipment gantry and patient table positioning information, in addition to all relevant technical information (e.g. peak kilovoltage (kVp), mA, ms, $K_{a,r}$, P_{KA}, copper (Cu) filtration).
(b) Display the reference air kerma rate ($\dot{K}_{a,r}$) (during imaging), cumulative $K_{a,r}$ for the procedure, and cumulative P_{KA} visible to the operator during the procedure. IEC also suggests that these systems provide an estimate of the patient maximum skin entrance dose and/or a skin dose map, both of which are becoming more widely available.
(c) Include last image hold (LIH) capability. This is an important dose saving feature that ideally should be routinely used clinically when live dynamic imaging is not necessary to visualize the desired clinical information. This feature displays the last complete image acquired from a radiation event until initiation of a new radiation event or the procedure is closed.
(d) Provide virtual centring, as it allows centring the patient on the last image without requiring use of radiation.
(e) Provide virtual representation of the collimator blade positions and wedge filters on the LIH, which allows the location of the collimator blades or wedge filters to be seen when they are adjusted, without requiring patient exposure.
(f) Include the ability to save a fluoroscopic imaging sequence, sometimes referred to as ‘fluoro save’. This feature avoids the need to acquire acquisition mode images only for archival purposes, thereby preventing unnecessary patient radiation dose. The ‘fluoro save’ feature is typically available only for a short time (~10s of seconds) after a given fluoroscopic event.
(g) Have the capability to disable radiation production is another important feature required by all applicable fluoroscopy equipment. This is typically a button located at the procedure table side controls as well as in the control room that disables the system to generate radiation. It is the best practice to always select this feature at the end of a case to ensure that there is no unintentional radiation exposure.

All equipment compliant with the IEC standard ought to have accompanying documents that are supplied by the fluoroscopy equipment manufacturer describing in detail the operational and functional aspects of the specific make and model of equipment. These documents are often referred to as the instructions for use. CQMPs ought to request and review these documents for all new fluoroscopic systems in order to familiarize themselves with the capabilities and radiation protection considerations unique to each system.

3.3. C-ARM GANTRY

The gantry that connects the X ray source and the image receptor (image intensifier or flat panel detector (FPD)) in an FGIP system is known as the C-arm because it has a semi-circular ‘C’ shape. The C-arm systems are divided into two types: fixed and mobile. The fixed C-arm system has the C-arm gantry permanently mounted on the ceiling (Fig. 8(a)) or floor (Fig. 8(b)), while the mobile system (Fig. 9) is portable and can be used in different rooms. The fixed systems are usually more sophisticated and connected to multiple monitor displays for more complex FGIPs, while the mobile systems are designed to be compact and portable for less complex procedures of relatively shorter duration.

The C-arm systems can be further divided into single plane and biplane systems. The single plane system consists of only one C-arm, whereas the biplane system consists of two C-arms (Fig. 10). Biplane angiography can potentially shorten the procedure time and reduce the amount of contrast agent injected.

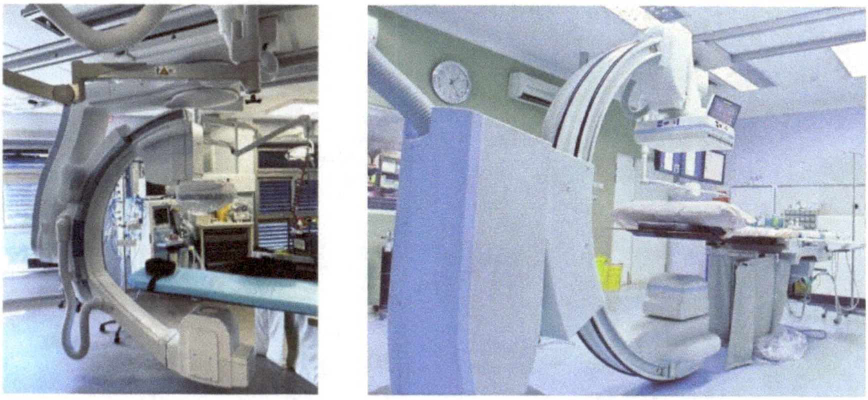

(a) (b)

FIG. 8. Examples of (a) ceiling-mounted and (b) floor-mounted C-arm fluoroscopy systems.

The C-arm can rotate over three axes, as shown in Fig. 11; rotation around the L-arm (R_z), rotation of the C-arm (R_y), and angulation of the C-arm (R_x).

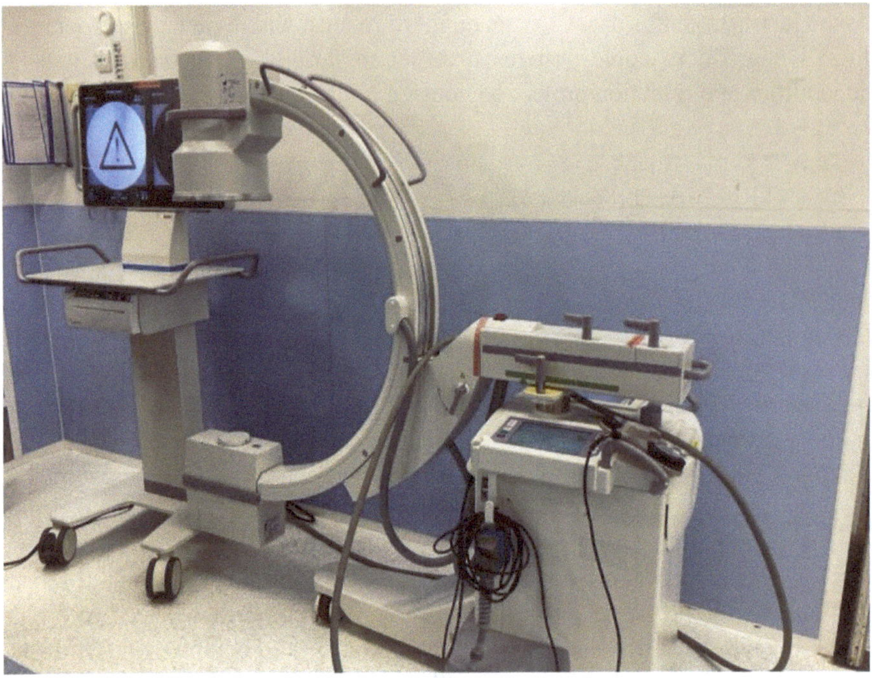

FIG. 9. A mobile C-arm fluoroscopy system.

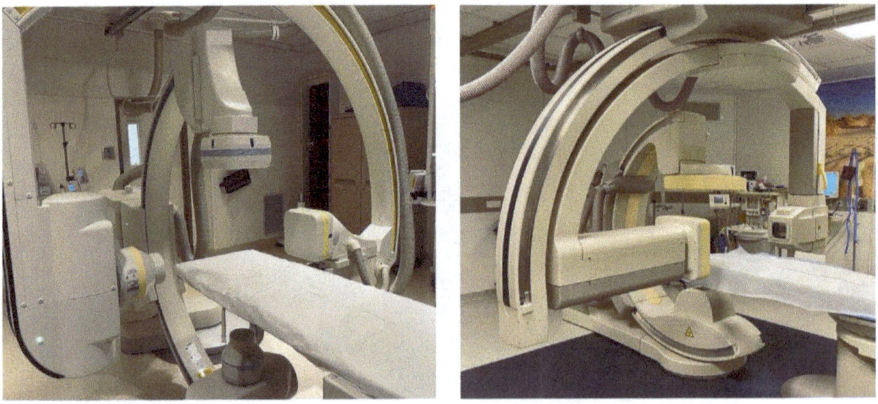

(a) (b)

FIG. 10. Examples of a biplane C-arm system for (a) cardiac and (b) neuro/body FGIPs.

There are some specific nomenclatures used to describe the orientation of the C-arm. First, the rotations within the plane of C-arm gantry are referred to as primary or orbital rotation (e.g. left-anterior-oblique (LAO) or right-anterior-oblique (RAO)) and denoted by an angle α (e.g. LAO 30°, RAO 30°, etc.). Second, rotations perpendicular to the plane of the C-arm gantry are referred to as secondary or angular rotations (e.g. cranial (CRA) or caudal (CAU)) and denoted by an angle β (e.g. CRA 60°, CAU 60°, etc.). The orientation angles and nomenclatures are illustrated in Fig. 12. Some robotic systems permit additional gantry movements, such as raising or lowering the gantry height in the vertical (ceiling-to-floor) direction (Fig. 13). Information regarding these gantry angles is typically included in the digital imaging and Communications in Medicine (DICOM) RDSR.

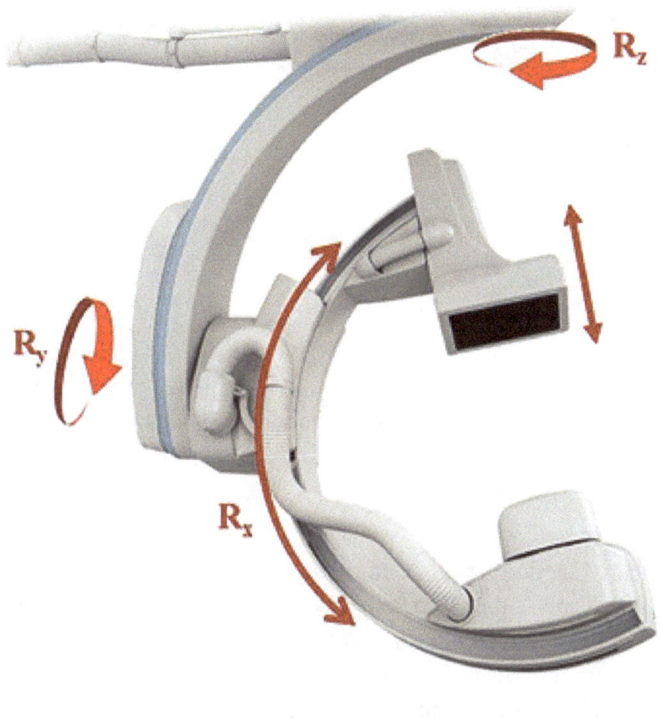

FIG. 11. Geometry of a C-arm gantry and its degree of freedom. R_x corresponds to the angulation of the C-arm and R_y to the rotation; R_z is the L-arm rotation as mentioned in Ref. [10].

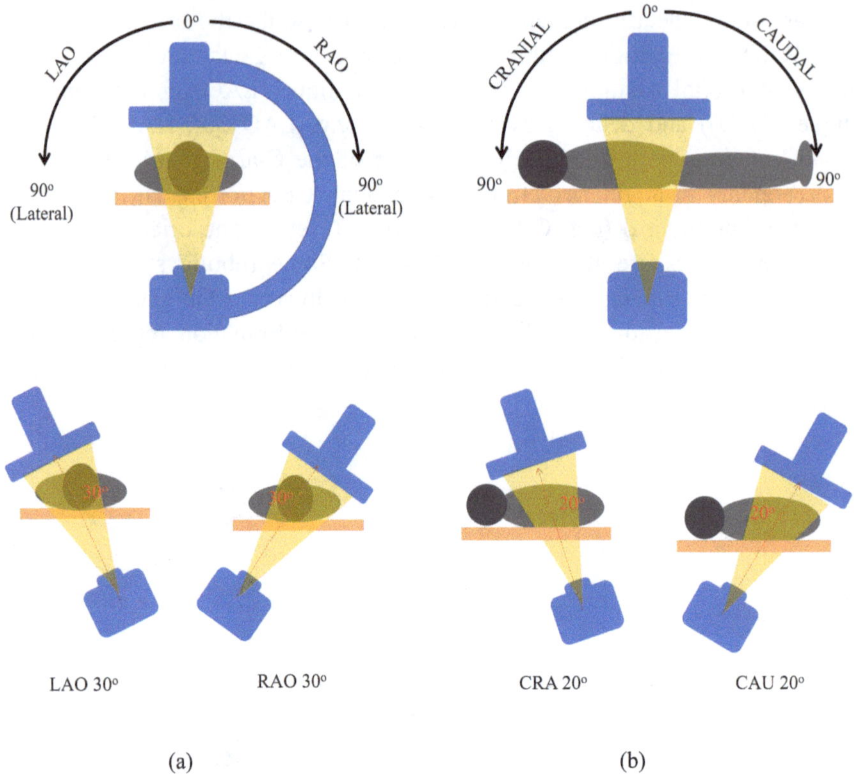

FIG. 12. Definition of the orientation angles and their nomenclatures in a C-arm gantry system: (a) primary or orbital rotation; (b) secondary or angular rotation.

3.4. X RAY GENERATOR

The main function of an X ray generator is to provide high voltage and current to the X ray tube. The generator power is typically in the range of 80–100 kW. For example, a high frequency converter or three phase twelve pulse generator is commonly used for fluoroscopy systems. All modern FGIP systems need to be capable of producing pulsed fluoroscopy with appropriate pulse rates and pulse widths for clinical tasks. Modern X ray generators are very robust and capable of sustaining high power load without overheating.

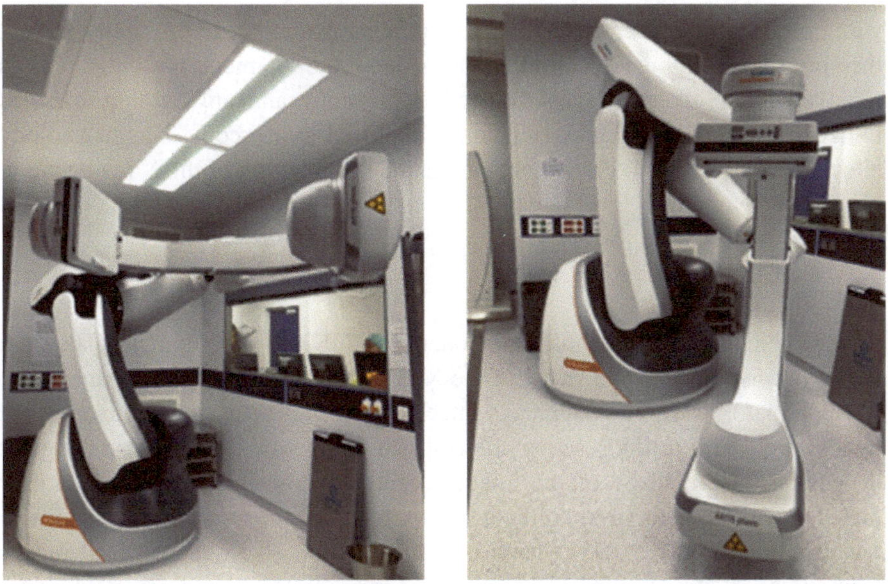

FIG. 13. Examples of gantry movements of a robotic system.

3.5. X RAY TUBE CHARACTERISTICS

The design and configuration of an X ray tube used in most fluoroscopy systems are similar to those used in general radiography. However, for special purpose rooms, such as those used for cardiac interventional procedures, extra heat capacity of the X ray tube is required in order to allow for prolonged exposure during complicated procedures and for the cases when higher dose radiographic images are acquired in rapid succession to visualise opacified vessels. The suggested heat capacity for a fluoroscopy system ranges from 0.4 to 1.35 mega heat units.

A typical X ray tube consists of the cathode, anode, rotor/stator, glass or metal evacuated envelope, tube port, cable sockets and tube housing, as shown in Fig. 14. The cathode consists of tungsten filament and a focusing cup. The filament is electrically connected to the filament circuit, which provides a voltage of approximately 10 V and variable current up to 7 A. Most fluoroscopy X ray tubes have two or three filament sizes, including large, small and potentially micro focal spot sizes, for detailed small vessel imaging such as in neurointerventional procedures. Special X ray tube designs, such as the grid-biased tube (also known as grid pulsed tube), may be used for pulsed fluoroscopy and cineangiography where rapid X ray pulsing is required. This type of X ray tube contains its own switch, which allows the X ray tube to be turned on and off rapidly, as is required with cine fluorography. A negative voltage bias is applied to the focusing cup, thereby containing the electron cloud around the filament until the bias is removed for a given pulse. Advanced fluoroscopy and angiography systems utilize grid-biased X ray tube switching for pulsed fluoroscopy. This technology allows for the rapid and precise activation and deactivation of the X ray beam.

3.6. COLLIMATOR ASSEMBLY

Collimators are essential components of all X ray equipment. They are used to adjust the size and shape of the X ray field emerging from the tube port. The collimator assembly is typically attached to the tube housing at the tube port by a swivel joint. In fluoroscopy, the collimation may be circular or rectangular depending on the shape of the image receptor. Image intensifiers are usually matched with circular (iris) collimation, while FPDs are typically matched with rectangular collimation. Two pairs of adjustable parallel-opposed lead blades (sometimes known as shutters) are used to form a rectangular X ray field, while an iris collimator is used to form a circular X ray field (Fig. 15).

For both the image intensifier and FPD fluoroscopy systems, the collimators need to automatically adjust to limit the X ray beam to the active

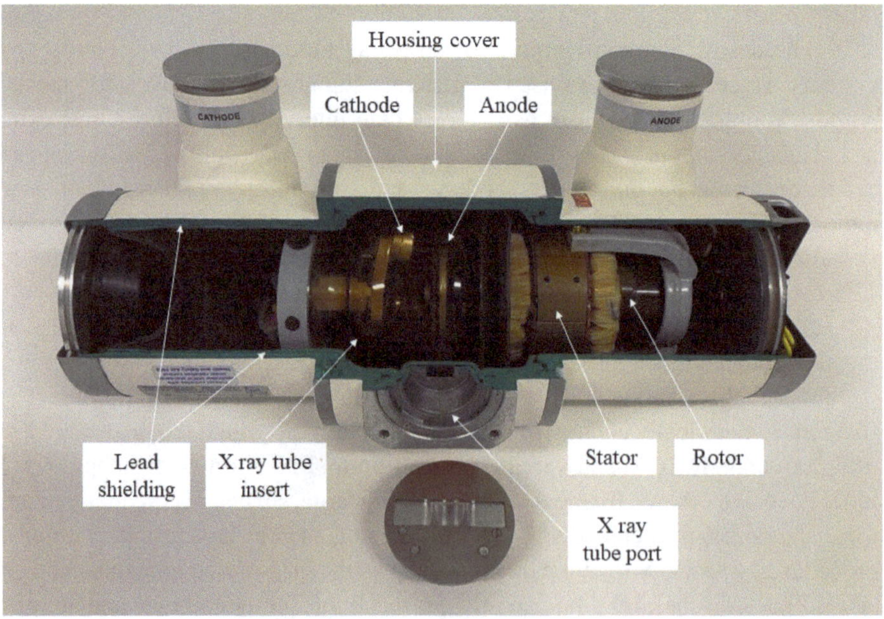

FIG. 14. The major components of a typical X ray tube (as outlined in Ref. [11]).

field of view (FOV) whenever the FOV or the source-to-image distance (SID), also called focal-to-image distance, is changed. Fluoroscopic systems used for FGIPs also contain equalization filters housed within the collimator assembly. These filters, also called contour or wedge filters, are partially radiolucent blades used to equalize the X ray flux across the image receptor when there are two or more very different attenuating structures (such as lung and mediastinum) present in the image.

Collimation is an effective method to reduce radiation exposure to the patients and staff while improving image quality by decreasing scattered radiation from the surrounding tissues, which degrades contrast. The goal of collimation is to irradiate only the region of immediate clinical interest. Obtaining a smaller FOV by selecting a magnification mode also reduces the X ray beam size and magnifies the displayed image, but most systems increase the dose rate when the FOV is reduced by this method.

3.7. FIXED AND SPECTRAL SHAPING FILTERS

Fluoroscopic systems commonly employ beam hardening filters to enhance X ray beam quality and minimize patient radiation dose. Within fluoroscopy, X ray beam filtration techniques can be broadly classified into two categories: fixed X ray filters and dynamic (or spectral shaping) filters.

Fixed X ray filters are permanently mounted inside the collimator assembly. Usually constructed of materials like aluminium or copper, they are strategically placed between the X ray source and the patient. Fixed filters absorb low energy

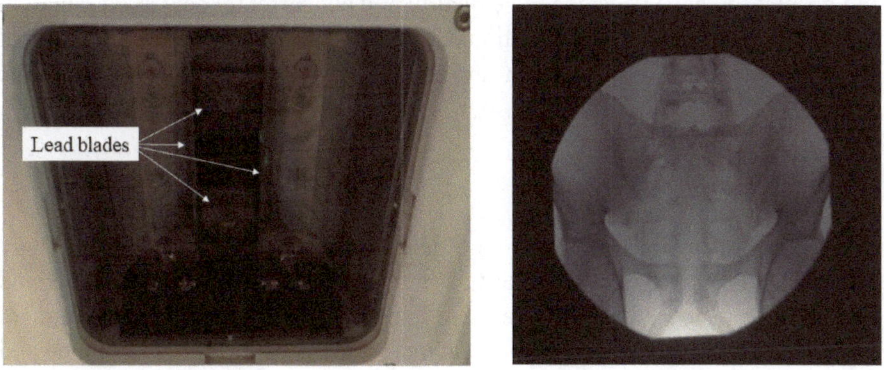

(a) (b)

FIG. 15. (a) Collimator assembly mounted on an X ray tube housing. (b) Example of iris collimation from an image intensifier fluoroscopy system.

X rays that are not valuable for image creation, as these would merely amplify the radiation dose to the patient without contributing to image quality. By eliminating these low energy X rays, fixed filters simultaneously upgrade the image quality and diminish radiation exposure to the patient.

Dynamic filters, or spectral shaping filters, are utilized to modify the quality of the X ray beam. Unlike fixed filters, dynamic filters are controlled by the automatic exposure rate control (AERC), allowing more precise tailoring of the beam. This type of filtration typically involves a wheel of thin layered materials of varied thicknesses that adjust the X ray beam on the basis of the amount of attenuation required.

Together, fixed and dynamic filters represent essential tools in modern fluoroscopy, enabling both image quality optimization and radiation safety. Fixed filters provide a baseline improvement in image quality and reduction in radiation, while dynamic filters offer the adaptability and precision required for specific clinical applications. The combination of these filtration techniques exemplifies approaches to imaging and safety through optimization of the X ray beam spectrum that nearly all modern vendors have employed.

3.8. IMAGE RECEPTORS

Since fluoroscopy was first developed, there have been considerable improvements in both the equipment and methodologies, which has influenced its clinical application. The advent of the X ray image intensifier and the television camera in the 1950s marked a significant advancement. Enhancements in intensifier technology and image displays, alongside progress in X ray tubes and generators, have boosted image quality and concurrently decreased radiation exposure to patients. This evolution has persisted with the advent of digital systems that utilize FPDs, which are now extensively available and continue to evolve [12].

3.8.1. Image intensifier

Image intensifiers represent an older analogue image receptor technology that is now used only with mobile C-arm systems. However, some older fixed angiography systems may still use these image receptors. Image intensifiers are comprised of an input screen, a photocathode, a focusing electrode, an anode and an output screen, which are housed in an evacuated assembly [13]. Figure 16 shows a diagram of the image intensifier.

X rays are converted into photoelectrons by a photocathode, and these electrons are then accelerated by a high voltage applied across the image intensifier. These electrons are focused by an electronic lens (guide), which consists of a focusing electrode and an anode, to project an image onto the output screen. In the input screen, the incident X ray photons are converted into light photons at a ratio of 1 to 3000. However, only 10 to 20% of these light photons are converted into electrons by the photocathode. The focusing electrodes guide the electrons to the output screen, where they are accelerated and undergo a magnification process. These accelerated electrons are then converted back into light photons in the output screen. The term 'gain' refers to the ratio of the luminance on the output screen to the entrance exposure rate, measured in units of candela per square metre per milligray per second ($cd \cdot m^{-2}/mGy \cdot s^{-1}$). This gain is a measure of the efficiency of the conversion process, reflecting how well the system translates incoming X ray photons into a visible image on the output screen.

The main disadvantages of an image intensifier system are loss of image contrast due to X ray and light scatter within the tube, the geometric distortion of the image due to the curved input phosphor, and an influence of earth's magnetic field or other ambient magnetic fields.

3.8.2. Flat panel detectors

There are two main types of FPDs, which are based on the following:

(a) Direct detection, which incorporates a photoconductor to produce electrical charges upon detection of an X ray. Direct detection FPDs use a

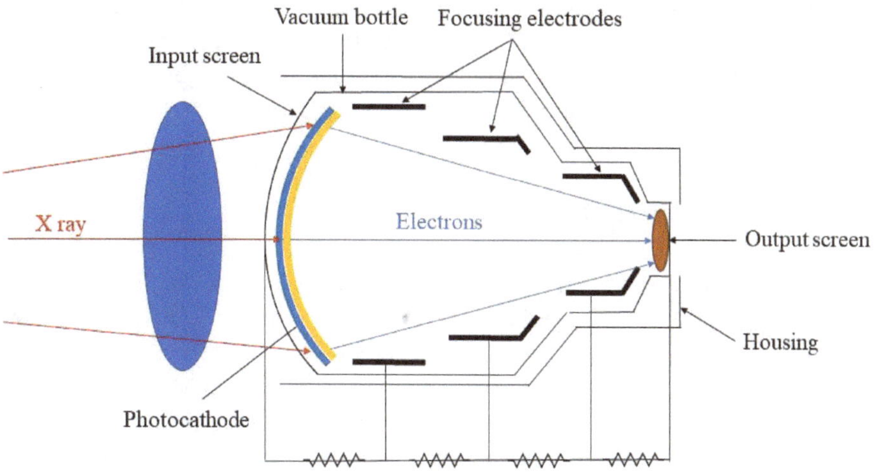

FIG. 16. Diagram of an image intensifier.

photoconductor, such as amorphous selenium, to directly convert incident X ray photons into electrical charges. This direct conversion process typically results in higher spatial resolution and better image detail, making these detectors particularly suitable for applications where fine detail is crucial, such as mammography and bone radiography.

(b) Indirect detection, which incorporates phosphor to produce visible photons upon detection of an X ray. Indirect detection FPDs employ a scintillator layer (commonly made of materials like caesium iodide or gadolinium oxysulfide) to convert X rays into visible light photons, which are then converted into electrical signals by photodiodes. These detectors are known for their superior dynamic imaging capabilities, crucial in procedures like angiography where capturing real time images of blood flow with minimal motion blur is essential. Additionally, they tend to be more cost effective and provide good quality images with good contrast, vital for differentiating blood vessels from surrounding tissues.

The vast majority of FPDs on the market are of the indirect detection type. The typical pixel pitch used for FGIPs ranges from 150 to 200 μm in the native pixel size. The choice between direct and indirect FPD will depend on the application. Indirect FPDs offer superior dynamic imaging capabilities, which is crucial in angiography for capturing real time images of blood flow with minimal motion blur. They also tend to be more cost effective and provide good quality images with good contrast, essential for distinguishing blood vessels from surrounding tissues. In FGIPs where a balance of image quality, dynamic range and cost is critical, indirect detectors frequently emerge as the preferred choice. The advantages of FPDs include freedom from electro-optics and magnetic field artefacts, better contrast, potentially lower dose, larger rectangular FOVs, a 'fluoro store' function (digital storage of real time fluoroscopic videos) and an easy choice of detector dose levels to suit operator's needs.

3.9. DISPLAY MONITORS

The display monitors used in FGIPs are a critical and indispensable component of the entire imaging chain, playing a vital role in the successful execution of these complex medical interventions. Classified as 'modality displays' according to American Association of Physicists in Medicine (AAPM) report 270, these high end displays are not only instrumental in conveying visual data but are also sometimes held to certain criteria from the 'diagnostic displays' classification, given their use in making clinical diagnoses [14].

FGIP theatres make use of various types and sizes of display monitors, adapting to specific clinical needs and procedures. Their positions within the theatre, such as next to the patient bed or inside the control room, are deliberately arranged to enhance workflow and facilitate effective patient care. Commonly, a large display monitor or a stack of display monitors is situated strategically for optimal viewing. While some displays are employed for navigational or administrative tasks, the primary displays that are used to present medical images are crucial in guiding the healthcare professionals through each intricate step of the intervention.

By presenting clear, accurate and real time images, these displays enable clinicians to precisely target and treat affected areas within the body, improving the efficiency, accuracy and safety of the procedures. The following paragraphs summarize the fundamental elements of display monitor technology and highlights their significance in supporting the common types of interventions carried out in FGIPs. They underscore the need for rigorous standards and considerations in selecting and utilizing these displays, reflecting their centrality in the sophisticated and life-saving care provided in FGIPs.

3.9.1. Cathode ray tube display

At the time of writing this report, the use of cathode ray tube (CRT) displays had been largely discontinued, and the description and testing methods specific to CRT technology is therefore not included. Readers are encouraged to refer to AAMP Task Group (TG) 18 Report Section 2.3.1 for a complete description of CRT technology, if required [15].

3.9.2. Liquid crystal display

Liquid crystal display (LCD) technology is currently the dominating technology used in FGIP displays and is described as an ‘transmissive’ type display technology, since the pixel matrix on these displays functions by altering the light intensity originating from a backlight source. In simple terms, the backlight source is filtered through a first polarizing filter and then moves through the pixel matrix, which is made up of an array of individual liquid crystals, before reaching the exit polarizing filter and the observer. The orientation of the liquid crystal molecules in relation to the polarizing filters is manipulated using small voltage changes in an external electric field. This, in turn, alters the opacity (or polarisation) of each liquid crystal cell, allowing varying intensities of light to pass through depending on the voltage applied. Thin film transistor array technology is used to apply the voltage changes over the liquid crystal cells so that the entire luminance range of the display between L_{min} and L_{max} can be shown.

3.9.3. Organic light-emitting diode (OLED) display

OLED technology is an emerging 'emissive' type display technology that differs considerably from LCD in that there is no backlight present, and each pixel element produces light through an electroluminescence process. Electroluminescence describes the direct conversion of electrical energy into light, which is achieved by accelerating a charge through an organic semiconductor or polymer material, which results in the production light though radiative recombination. The OLED luminance response, or light output, is controlled by applying specific current levels to each individual pixel in the pixel matrix.

OLED displays offer excellent contrast and high L_{max} values and require less energy to produce than LCD due to the direct electroluminescence process and absence of a backlight. OLED is a developing technology in medical display devices and still suffers from differential ageing of materials, luminance degrading with time and residual image artefacts [16]. This publication focuses on the testing of LCD displays, and users need to be cautioned that testing techniques may need adaption for advancements in OLED displays in the future.

3.10. PATIENT TABLE

For fixed C-arm equipment, a patient support table and mattress are integrated into the imaging system. These supports ideally should be able to hold a large amount of weight and withstand the potential need for cardiopulmonary resuscitation on a patient, given their interventional nature. In addition, the table also needs to be radiotranslucent to avoid attenuating the X ray beam as much as reasonably possible. Likewise, the patient mattress needs to be comfortable for the patient while being reasonably radio translucent. In general, the tables and mattresses provided by manufacturers will likely be optimal for X ray imaging procedures. The attenuation properties of all tables and mattresses used during FGIPs need to be evaluated during acceptance testing and whenever they are replaced or modified. Further discussion on performing this type of evaluation can be found in Section 5.2.20.

3.11. AUTOMATIC EXPOSURE RATE CONTROL

AERC, previously known as automatic brightness control for image intensifier technology, is a control system/logic used to automatically modulate X ray exposure parameters (kVp, mA, pulse width, filtration) based on the thickness and density of the anatomy being imaged in order to maintain a consistent exposure at the image receptor as efficiently as possible [11]. How the system determines the most efficient means of adjusting the technique factors is largely based on the selected exam protocol and the background settings that are chosen within that protocol. For digital image receptors, the AERC uses the X ray signal from a pre-selected area of the image receptor to determine the necessary X ray output and beam quality. The size and shape of the field used for this purpose is often selectable for a given imaging protocol or exam and modifiable by the user. For many FGIP systems, collimating within this area may cause unexpected or unpredictable results.

For signal-to-noise ratio (SNR) targeted systems, the AERC circuitry attempts to keep the detected photon fluence in each fluoroscopic frame constant so that the SNR of the image remains constant regardless of the patient's thickness or density. It also increases the exposure rate when magnification mode is used, typically inversely proportional to the area of the X ray beam. The AERC regulates the kV, mA, ms and potentially spectral filtration in a predetermined manner, as defined by the chosen protocol. However, how these parameters vary depends on the model of FGIP system and even among the imaging protocols on a given system. Most fluoroscopy systems provide the options of 'low', 'normal' and 'high' fluoroscopy selections on the console. When the 'high' mode is selected, the AERC will typically respond by increasing the mA instead of the kV to preserve subject contrast but at the expense of higher patient dose. On the other hand, when the 'low' mode is selected, the AERC will typically respond by first increasing the kV to increase photon transmission but decreasing subject contrast [11].

3.12. OPERATION MODES

This section discusses the multiple operation modes used in FGIPs. Each of the operation modes has its own advantages and limitations, and the choice of mode depends on the specific needs of the procedure. The radiological medical practitioners who perform the procedure will determine which mode to be used on the basis of their clinical judgment and the patient's condition. There are two primary modes of operation: fluoroscopy mode and acquisition mode.

3.12.1. Continuous fluoroscopy mode

In modern fixed FGIP systems, continuous fluoroscopy is no longer commonly used or even available. However, it is still found on many mobile C-arm devices. When imaging in this mode, the X ray beam is continuously active, while the system acquires images typically with a frame rate of 30 frames per second (fps) [11]. While this mode of fluoroscopic imaging can provide high temporal resolution, allowing for detailed real time imaging, it can also result in a higher radiation dose to both the patient and staff. This contrasts with pulsed fluoroscopic imaging, which tends to reduce radiation exposure [17].

3.12.2. Pulsed fluoroscopy mode

The pulsed fluoroscopic imaging mode, almost universally found in modern interventional fluoroscopes, helps reduce radiation dose rates to both patients and staff compared with continuous fluoroscopic imaging. This is achieved by delivering X rays in brief, clearly defined pulses rather than continuously. Short pulse widths can often enhance image quality by reducing exposure time and minimizing motion blur. Interventional fluoroscopy systems generally offer a wide range of pulse rates, from as low as 0.5 pulses per second (pps) to 30 pps. Since patient entrance air kerma is related to the pulse rate, clinicians need to opt for the lowest pulse rate that meets the specific temporal requirements of the clinical procedure [18]. It needs to be noted that different manufacturers use varying scaling factors when adjusting pulse rates. This variation stems from the distinct design and calibration protocols the manufacturers employ in each of their fluoroscopic systems. Wherever possible, shorter pulse widths minimize image blurring from patient or anatomical movement, although the consequence is the need for a higher mA to maintain the same current–time product (mAs) (and dose) per pulse. Consequently, pulsed fluoroscopy usually yields better image quality at reduced dose rates compared with continuous fluoroscopy. This advantage explains why most modern interventional fluoroscopes are equipped exclusively with the pulsed option [11].

3.12.3. Acquisition/cine mode

The acquisition mode, also called cine mode, plays a pivotal role in capturing good quality images necessary for diagnosing and treating vascular conditions. During acquisition, the machine switches to a higher dose and resolution setting compared with the fluoroscopy mode, enabling it to capture finer details essential for accurate vessel visualization. The acquisition mode typically operates at higher frame rates and utilizes advanced image processing techniques to enhance clarity and reduce motion blur, which are crucial for precisely delineating the anatomy and pathology within the vascular system. The images obtained in this mode are often used for archiving purposes as well as for planning and guiding FGIPs. The dose in cine mode is directly related to the frame rate — higher frame rates result in higher doses. Frame rates can range from 1 to 30 fps or even higher, depending on the clinical requirement type of patient (adult or paediatric), etc. [3, 19].

3.12.4. Magnification mode

Magnification mode is available in most fluoroscopy systems to allow for a magnified view of the clinical area of interest. Magnification mode works by using a smaller FOV to capture the image. On an image intensifier system, the size of the X ray at the input phosphor can be varied electronically for the same output phosphor size [20]. When the magnification mode is engaged, the X ray beam collimator will automatically adjust to match the X ray beam dimensions to the smaller FOV. The smaller FOV increases the spatial resolution, resulting in a clearer image. The magnification factor can range from 1.5 to 4 times the normal size, depending on the specific system used (Fig. 17). The brightness gain (and hence the overall gain) of the image intensifier decreases as the magnification increases. The AERC circuitry will then increase the X ray exposure rate to compensate for the lower signals.

FPD systems typically exhibit a relationship between magnification factor and exposure rate; however, this relationship differs from that in image intensifier systems. Unlike image intensifiers, FPDs maintain contrast at lower dose rates due to their linear response range. However, a key consideration in FPD systems is the management of apparent noise. While contrast can be maintained, the SNR may not be constant across varying dose rates, leading to an increase in apparent noise, particularly at lower doses. This factor plays a critical role in image quality and needs to be carefully managed, especially when adjusting magnification and exposure rates in FPD systems. When magnification mode is activated in FPD systems, the X ray beam collimator adjusts to align the X ray beam with the smaller FOV. This reduced FOV enhances spatial resolution and decreases image noise, thus improving image quality. Magnification factors typically range from 1.5 to 4 times the normal size. When a smaller FOV is selected in FPD systems, there is often a practice of 'binning' digital elements. Binning involves combining the signals from multiple digital elements to behave as a larger single detector element. This process can improve SNR but at the cost of reduced spatial resolution. However, it also influences the dose rate; as fewer digital elements are used in the magnified mode, the radiation dose per digital element increases to maintain image quality. Therefore, while magnification provides clearer

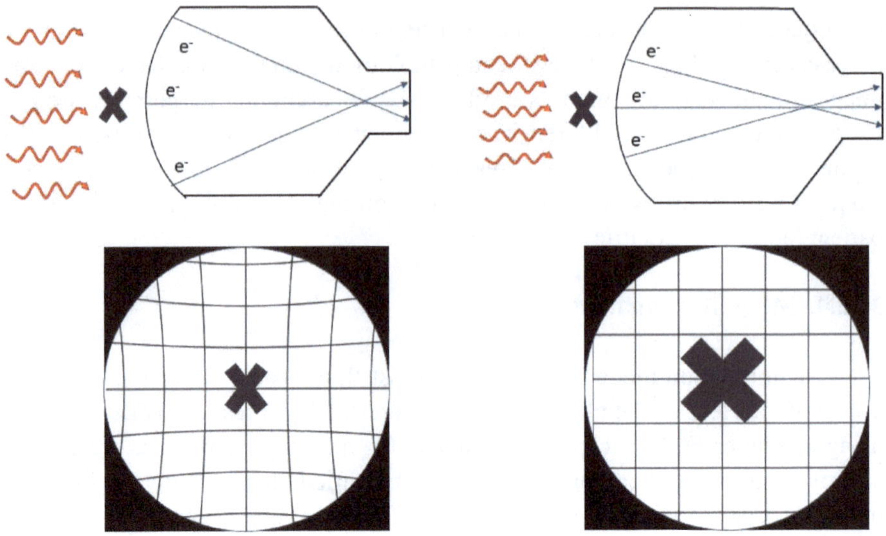

(a) (b)

FIG. 17. (a) Normal mode (no magnification). (b) In magnification mode (smaller FOV) for image intensifier, objects are magnified, and spatial resolution is improved, but the brightness gain of the image intensifier is decreased, and hence higher X ray exposure rate is required to compensate for the lower signals. Note that the pincushion distortion is also reduced in the magnification mode, as there is less warping onto the output phosphor from the smaller central area of the input phosphor.

images of small details, which is essential in many diagnostic and interventional procedures, it raises considerations regarding patient dose. The increased dose in smaller FOVs due to the higher exposure required for maintaining image quality necessitates careful use of magnification. Collimation instead of magnification (reducing input FOV) ideally should be used if the user intends to view smaller region of interest.

3.12.5. Digital subtraction angiography (DSA) mode

The fluoroscopic imaging mode known as DSA employs a vascular contrast agent, typically iodine, to augment the visibility of blood vessels or other internal structures. In DSA, X ray images are captured both before and after the injection of the contrast agent. The image taken prior to the contrast, referred to as the mask image, displays the anatomy in the image volume. Successive images taken post-contrast illustrate the contrast-enhanced vessels overlaid on the background anatomy.

To better visualize the contrast-filled vessels, the mask image is subtracted from the post-contrast images. The resultant subtracted images clearly depict the contrast-enhanced vessels or tissue while excluding the background anatomical structures (see Fig. 18). Relative to non-subtracted imaging techniques, DSA can enable lower contrast agent doses and provide clearer images of contrast-enhanced anatomy, provided that there is minimal anatomic or patient motion, which can cause considerable artefacts in the image due to misregistration during the subtraction process. However, it is worth noting that the process of image subtraction can reduce the SNR, causing the subtracted images to appear noisier. To mitigate this effect, increased X ray dose rates are used to maintain sufficient SNR. This is the reason why DSA is considered one of the highest dose rate modes of imaging, though this mode of imaging is invaluable for body and neuro interventions providing unparalleled clinical information in both 2-D and even 3-D modes on some systems.

3.12.6. 3-D rotational acquisition and cone beam CT mode

This mode is used to create 3-D images of the anatomical structure. The C-arm is rotated around the patient and captures projection images from various angles. The images are then reconstructed into a 3-D image dataset and can be evaluated or overlaid on live fluoroscopic images to aid in anatomic guidance or device placement. Some systems only reconstruct 3-D rotational scans while others perform more CT-like reconstruction using filtered back projection or iterative reconstruction methods used in traditional CT image formation. One of the advantages of this mode is that it can produce 3-D images with a lower radiation dose compared with conventional CT scans and typically better

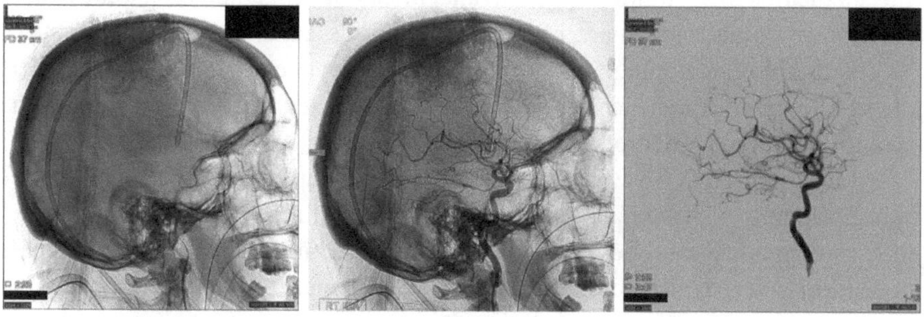

(a) (b) (c)

FIG. 18. Basic principles of DSA. (a) a mask image is taken before contrast injection; (b) a contrasted image; and (c) DSA image where the mask image is subtracted from the contrasted image pixel by pixel, hence showing only the contrast-filled vessels.

spatial resolution [21]. However, image contrast, especially for soft tissue, is of substantially lower quality than conventional CT scans due to the large amount of scatter produced from the volume of anatomy imaged during a rotation. Cone beam computed tomography (CBCT) with full image reconstruction or simple 3-D rotational imaging are increasingly used in FGIPs. They are used in some clinical specialties such as cardiac imaging for normal coronary arteriography where two 3-D passes can yield the desired results for the entire procedure. In neurointerventional cases, it has been shown that full CBCT provides more information than 3-D angiography and is generally required for complex cases like the management of aneurysms. In the neurology field, HR-CBCT improves the understanding of diseases and guides therapeutic decisions. Both 3-D rotational imaging and CBCT are used in chemoembolization of hepatocellular carcinomas, helping in detecting tumour shape and tumour feeders.

3.13. LAST IMAGE HOLD AND LAST SEQUENCE DISPLAY

LIH is a vital and standard feature incorporated into all modern fluoroscopy systems, reflecting a significant advancement that reduces the radiation dose to the patient and staff. When the clinician releases the 'fluoro' pedal, or footswitch, the system does not simply revert to a blank monitor. Instead, the last image captured is frozen and displayed on the screen.

This functionality serves multiple purposes, such as allowing for continued analysis. The held image enables the clinician to examine the specific area or structure in detail for an extended period without the need for continuous radiation exposure. This provides a more extended window for critical assessment without any additional radiation risk, fostering a safer environment for both patients and medical staff. Additionally, advancements in imaging technology now incorporate virtual movement capabilities, such as adjustments in collimation, FOV and table positioning. These virtual adjustments can be made on the LIH image, allowing for enhanced visualization and focused examination without the need for additional radiation exposure.

Most modern FGIP systems are equipped with larger video buffer memories and additionally provide the option to replay the most recent sequence of images. This functionality can be particularly valuable during complex or delicate procedures where the radiological medical practitioner might need to review the recent sequence of events to ensure that the procedure is progressing as expected, or to make informed decisions about the next steps in the intervention.

The LIH feature in modern fluoroscopy systems is more than a simple convenience; it is a multifaceted tool that contributes to improved safety, efficiency, collaboration and patient care in the dynamic and demanding context of interventional procedures. Its integration into the standard capabilities of fluoroscopic devices underscores its value in contemporary medical practice.

3.14. FRAME AVERAGING (RECURSIVE FILTRATION)

Fluoroscopy provides excellent temporal resolution, but the images are relatively noisy in order to minimize patient dose. To overcome this limitation, image frames are summed to increase the SNR of the displayed image. In this technique, multiple consecutive frames are captured and summed together in a weighted fashion, which increases SNR [22]. The more frames averaged together, the greater the reduction in noise and the smoother the resulting image. However, there is a trade-off; increasing the frame averaging also integrates frames over a longer period of time, and any motion in the image will substantially degrade the resulting displayed image. Different temporal averaging algorithms are available, but a common one is called recursive filtering. In this approach, the newly acquired image, I_n, is added to the last displayed image (I_{n-1}) using the following equation [11]:

$$I_{\text{displayed}} = \alpha I_n + (1-\alpha)\, I_{n-1} \tag{1}$$

where α is a weighting factor and ranges from 0 to 1.

As α is reduced, the contribution of the current image (I_n) is reduced, the contribution of the previous displayed image is enhanced, and the amount of lag is increased. As α is increased, the amount of lag is reduced, and at $\alpha = 1$, no lag occurs, but with no frame averaging and hence lower SNR. The current displayed image ($I_{\text{displayed}}$) becomes the I_{n-1} image for the next displayed frame, and the weighted image data from a predefined number of images are included. This setting is found within the background parameters for each fluoroscopic mode of imaging within each imaging protocol. In some fluoroscopy systems, recursive filtering is applied selectively. For instance, it can be regionally disabled in areas of the image where high temporal resolution is critical, such as around contrast-filled vessels, to preserve the clarity of these structures. In other parts of the image, where less detail is required or where motion is minimal, recursive filtering is applied to enhance SNR. This selective application optimizes image quality, ensuring that important details are captured clearly while still maintaining overall image smoothness and reducing noise in less critical areas.

3.15. ROAD MAPPING

This is a software and video enhanced variant of the LIH and DSA features to track advancement and placement of devices (such as catheters) within the body. The process starts with a contrast injected image or DSA image, inverting the grayscale, and then overlaying that inverted image on the live fluoroscopy image. By doing this, the radiological medical practitioner has a 'roadmap' of the patient's vascular anatomy superimposed on the fluoroscopic image and contrast inverted to better visualise the catheter or guidewire through the vasculature [22]. Road mapping is especially beneficial when advancing through complex and tortuous vessels (see Fig. 19).

3.16. CLINICAL IMAGING PROTOCOLS

An imaging protocol is a set of parameters that include default dose rate levels (e.g. low, medium, high, etc.) or image receptor dose per pulse (e.g. low, medium, high, etc.), continuous or pulsed fluoroscopy, pulse or frame rate (e.g. 0.5–30 pps or fps), pulse length (in ms), kV-mA curve, as well as many image processing parameters. The operator can normally choose from a series of patient protocols that are tailored to the clinical application and patient body habitus.

In normal practice, protocols are usually defined by the manufacturer, though they may be altered by the end users. They ideally should be altered only

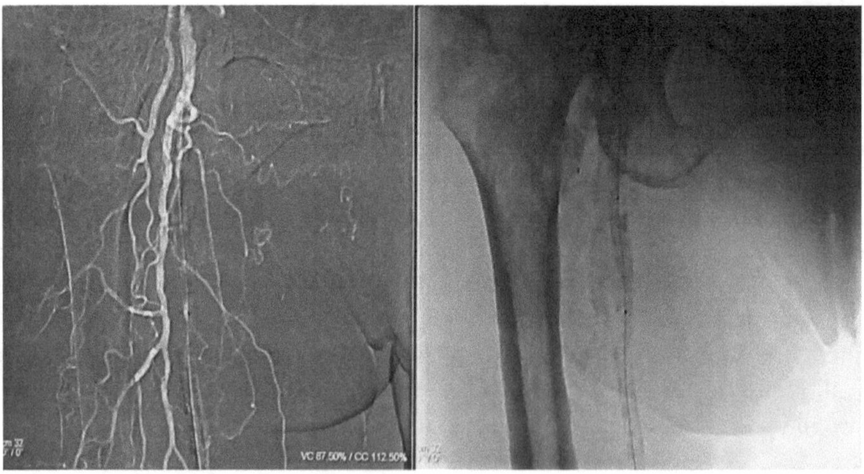

FIG. 19. Examples of road mapping. Image on the left provides a roadmap overlaid on the current fluoroscopy image acquired from a DSA image.

with collaborative input from the vendor, clinical team and the CQMP. It is worth noting that the protocol labels may not always be representative of or optimized for the specific imaging tasks. For example, cardiac imaging characteristics typically include short pulse length, and pulse or frame rates. More specifically, additional considerations may include the following:

(a) For coronary angiography: low and moderately difficult percutaneous coronary interventions versus difficult percutaneous coronary interventions, low heartbeat rate (<65 bps) versus high heartbeat rate (>70 bps), thin, normal and obese patients (body mass index (BMI) < 20, $21 < \text{BMI} < 30$, $\text{BMI} > 30$ kg/m^2).
(b) For cardiac electrophysiology device implantation or ablations: thin, normal and obese patients.
(c) For transcatheter aortic valve implantation: thin, normal and obese patients.

For neurointerventional imaging of the neurovasculature, the brain pathology, sensory organs, meninges, cerebrospinal fluid, head and neck, spinal cord, vertebral column and the peripheral nervous system typically do not require as broad a range of protocols as other clinical disciplines.

Mobile C-arm systems may have only one or two kV-mA curves, while some systems offer various imaging protocols tailored to specific clinical applications, such as urological, orthopaedic, cardiac and vascular procedures. These curves represent predefined settings of kV and mA values that the system uses to achieve the desired image brightness and quality. This is particularly important in varying clinical scenarios, where the thickness and density of the imaged body part can significantly differ. The limited number of kV-mA curves in some mobile C-arm systems may restrict their versatility across different types of procedures, whereas systems with a broader range of curves can more precisely optimize imaging parameters for a variety of clinical applications.

3.17. DOSE INDICATORS

This section outlines the principal radiation quantities and units, which are essential in quantifying patient dose during FGIPs. It provides an in-depth understanding of key dose metrics that play a vital role in assessing radiation exposure. Furthermore, this section elaborates on how these measurements are utilized in clinical practice, aiding in the monitoring and management of patient doses.

3.17.1. Air kerma

The air kerma (K) for ionizing uncharged particles is the quotient of dε_{tr} by dm, where dε_{tr} is the mean sum of the initial kinetic energies of all the charged particles liberated in a mass dm of a material by the uncharged particles incident on dm, thus [23]:

$$K = \frac{\mathrm{d}\varepsilon_{tr}}{\mathrm{d}m} \tag{2}$$

Unit: J/kg or Gy.

3.17.2. Air kerma rate

The air kerma rate ($\dot{K}$), is the quotient of dK by dt, where dK is the increment of K in the time interval dt, thus [23]:

$$\dot{K} = \frac{\mathrm{d}K}{\mathrm{d}t} \tag{3}$$

Unit: J/kg·s or Gy/s.

3.17.3. Entrance surface air kerma

The entrance surface air kerma ($K_{a,e}$), often referred in the literature as (ESAK), is the K on the central X ray beam axis at the point where the X ray beam enters the patient or phantom. The contribution of backscattered radiation is included.

Unit: J/kg or Gy.

$K_{a,e}$ is related to the incident air kerma ($K_{a,i}$) by the backscatter factor (B), thus:

$$K_{a,e} = K_{a,1}B \tag{4}$$

The backscatter factor depends on the X ray spectrum, the X ray field size, and the thickness and composition of the patient or phantom [24].

3.17.4. Entrance surface air kerma rate

The entrance surface air kerma rate ($\dot{K}_{a,e}$) is the quotient of d$K_{a,e}$ by dt, where d$K_{a,e}$ is the increment of entrance surface air kerma in the time interval dt, thus [24]:

$$\dot{K}_{a,e} = \frac{dK_{a,e}}{dt} \tag{5}$$

Unit: J/kg·s or Gy/s.

3.17.5. P_{KA}

The P_{KA} is the integral of the air kerma free-in-air over the area, A, of the X ray beam in a plane perpendicular to the beam axis. Older terminology for this quantity is 'dose area product' or 'kerma area product'.

Unit: (J/kg)·m^2 or Gy.cm^2.

3.17.6. Air kerma at the patient entrance reference point

The air kerma at the patient entrance reference point ($K_{a,r}$) is the position at which the displayed cumulative air kerma for FGIPs is determined as an approximation of the air kerma incident on the patient's skin surface. According to the IEC [9], this point is defined as lying on the central axis of the X ray beam, 15 cm from the isocentre in the direction of the focal spot. The location of the reference point relative to the X ray gantry does not change when the SID is changed. This quantity is referred to in older publications as 'cumulative dose' and has also been called 'reference air kerma' and 'reference point air kerma,' which does not include backscatter. It is important to understand that this is a fixed location in space and may or may not represent the actual X ray beam entrance on a patient for a given irradiation event.

Unit: J/kg or Gy.

3.17.7. Peak skin dose (PSD)

The maximum absorbed dose to a localized region of skin is referred to as PSD or maximum skin entrance dose and is measured in Gy [2, 24]. The methods to obtain PSD include the following:

(a) Direct measurement: The most straightforward methods for measuring skin dose are by the use of thermoluminescent dosimeters or optically stimulated luminescent dosimeters, scintillation dosimeters, or radiochromic film. Semiconductor dosimeters, such as diodes or metal oxide semiconductor field effect transistors, are typically used only for phantom measurements [24].

(b) Calculation: Ideally the calculation can be based on radiation event data from the DICOM RDSR: The DICOM RDSR is an information object that contains data from radiological procedures in a standard format. An RDSR contains structured information organised in a hierarchical tree structure [25, 26]. A vast amount of information is stored in the RDSR, including radiation dose quantities (procedure and radiation event level), patient related information, procedure information, and image acquisition technical information, in addition to private vendor information [27, 28]. RDSR availability on equipment used for FGIPs was mandated by IEC in 2010 and needs to be available on all subsequently manufactured equipment [9]. It is important to note that, before the implementation of the RDSR, many systems captured an image of the dose report screen, detailing various technical parameters. While these reports held less information compared to the RDSR, and the data typically could not be processed directly, they were still viewable by humans. Additionally, there were both commercial and open source software options available that could perform optical character recognition and text parsing to extract values and partially fill an RDSR or other databases with the retrievable information.

Details on PSD calculation and patient dose assessment are provided in Appendix III.

3.18. ARCHIVING SYSTEMS

This section describes the basic features of picture archiving and communication systems (PACS) and dose archiving and communication systems (DACS) used in FGIPs.

3.18.1. PACS

PACS is a combination of networked computer systems and other interfacing devices that processes information received from the various types of imaging modalities. The PACS then controls the image processing, display devices, picture storage, database features, backup, distribution and networking aspects of the radiology images. PACS needs to be able to accept DICOM standard images from the FGIP equipment, which is a standard format. Standard DICOM headers and metadata enable PACS to interface with larger radiology information systems or hospital information systems and correctly link patient images and data. This function is very important for FGIPs, as there are many clinical specialists using the equipment, which requires access to the patient images throughout the hospital network. A well-functioning PACS system that is sufficiently supported has become an essential component for modern healthcare facilities to operate effectively. It is noteworthy that the integration of RDSRs within PACS is varied across different systems. While some PACS are fully equipped to accept and process RDSRs and provide detailed dose information for each procedure, others may have limitations in handling this data or may not support RDSRs at all. Moreover, challenges arise in scenarios where a procedure results in zero saved images, a situation not uncommon in certain FGIPs. In such cases, the absence of image data can complicate the storage and retrieval of procedure related information, including dose metrics, within the PACS. This issue underscores the need for advanced PACS solutions capable of accommodating a range of data types and scenarios, ensuring complete and accurate documentation of all aspects of patient care, including those procedures where image saving is minimal or non-existent.

3.18.2. DACS

DACS is specifically designed to aid the clinical user in the comprehensive management of radiation dose index data, encompassing the collection, aggregation, analysis and archiving of this crucial information. These systems play a pivotal role in helping healthcare facilities establish local diagnostic reference levels (DRLs) or trigger levels, which are instrumental in identifying radiation exposure levels that necessitate clinical review or intervention. The development and implementation of DACS are in response to various guidelines and directives aimed at enhancing the management of patient doses, particularly in FGIPs [27]. Moreover, the integration of DACS with X ray systems facilitates the export of RDSR, enhancing the documentation and analysis of patient exposure data.

DACS serves as a crucial bridge in terms of interfacing with fluoroscopy systems and PACS. When integrated with fluoroscopy systems, DACS directly retrieves and processes radiation dose information generated during procedures. This seamless connection ensures that accurate and real time dose data are captured, allowing for immediate assessment and adjustment if necessary. The ability of DACS to interface with various fluoroscopy systems, regardless of the manufacturer, is key to its effectiveness in diverse clinical settings. Furthermore, the integration of DACS with PACS systems is of significant importance. This integration allows for the efficient transfer and storage of dose data within the larger imaging and patient health information framework of a healthcare facility. The dose information collected and processed by DACS, including RDSRs, can be archived within PACS, making it accessible for long term analysis, quality assurance and regulatory compliance. This connectivity ensures that radiation dose data are not only collected but also effectively incorporated into the patient's overall imaging record, allowing for a holistic approach to patient care and radiation safety management.

3.18.3. Dose management systems

In recent years, dose management systems have become critical in the field of medical imaging, playing a key role in quality assurance, patient dose management and optimization [29]. These advanced systems gather, monitor and evaluate a wide range of patient data, including demographic and complex technical details, as well as dose information from various imaging modalities. They are essential in improving quality assurance programs and complying with strict regulatory requirements. Studies also underscore their importance in optimization processes, ensuring the balance between image quality and appropriate radiation levels. Dose management systems offer a broad range of functionalities beyond just collecting data related to patient doses [30]. They efficiently extract a vast array of additional data, which significantly enhances patient care and supports continuous quality improvement efforts. The effectiveness and applicability of these systems vary on the basis of their unique features and capabilities, encompassing both high end commercial offerings and flexible open source options. A key advantage of dose management systems is their ability to automate processes like data collection, storage, analysis and reporting. This automation marks a significant advancement over older methods that depended on manual or semi-automated processes. Such automation not only makes workflows more efficient but also improves the precision and dependability of data handling in radiological practices.

3.19. ANCILLARY EQUIPMENT

This part describes some of the ancillary equipment commonly used in FGIP facilities. Examples of these systems are the footswitch or fluoroscopy pedal, the automatic contrast agent injector, the radiation protection equipment as well as the uninterrupted power supply (UPS).

3.19.1. Footswitch

FGIP equipment is generally equipped with a footswitch controller that gives the radiological medical practitioners control of the fluoroscopic and other functions of the machine using their foot only. This is required because the radiological medical practitioners would need to use their hands to control the movement of catheters and other devices while performing the procedure. There are different types of footswitches available to accommodate different user preferences and comfort. Figure 20 shows two examples of basic footswitches used on both single and biplane interventional machines. Figure 21 shows an example of a larger footswitch that also offers the user more advanced functions. These types of footswitches have a guard to prevent accidental actuation. They can be battery operated, so there is no need for cabling for ease of use.

3.19.2. Automatic contrast medium injector

FGIP techniques such as DSA require image sets with and without contrast medium to form the subtracted image. In DSA cases, contrast medium is injected into blood vessels, and the precise timing and position of the injection in relation to the radiographic image acquisition is important for optimal imaging. Different types of automatic contrast injector pumps are used for this purpose, and pumps can usually be linked to the interventional equipment to ensure accurate and consistent results. Figure 22 shows an example of an automatic injection pump system attached to the bed of an FGIP unit.

3.19.3. Radiation protection equipment

FGIP equipment uses X ray exposure extensively during clinical procedures, which makes radiation protection considerations and radiation

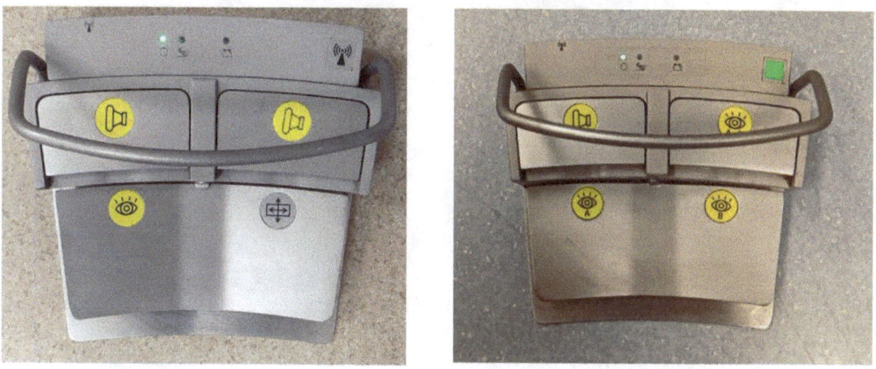

(a) (b)

FIG. 20. (a) Standard single plane interventional unit footswitch and (b) standard biplane interventional unit footswitch.

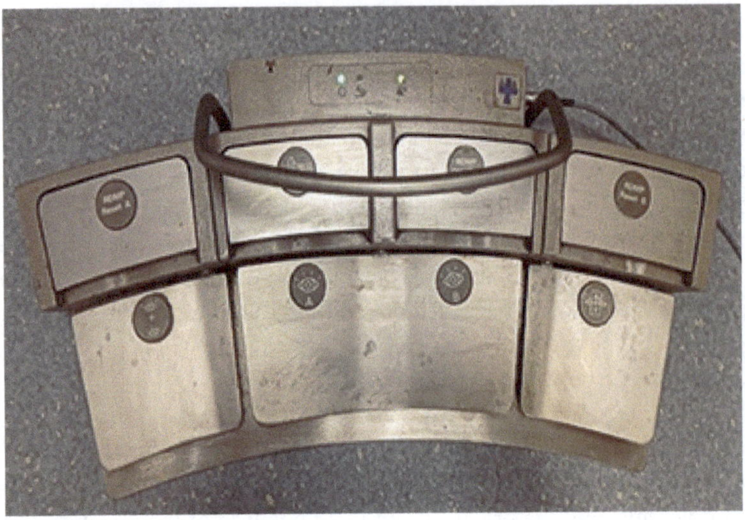

FIG. 21. Example of an advanced biplane interventional unit footswitch including functions for table movements and reference image controls.

FIG. 22. Example of an automatic contrast agent injector attached to the table of an FGIP unit.

protection equipment very important to ensure safety for the radiation workers. The design of the facility's structural shielding, personal protective devices, personal radiation monitoring devices and operational controls need to be considered when designing and equipping an FGIP facility. Common personal protective equipment used in FGIPs includes lead or lead-free aprons of appropriate type and thickness, thyroid shields, protective eyewear and various protective shields, including table drapes, a ceiling suspended protective shield, and mobile protective shields or drapes. The detailed specifications of radiation protection requirements are beyond the scope of this publication.

3.19.4. UPS

FGIP equipment should ideally be supplied with electricity backup systems in the form of non-interruptible power supply systems to prevent the equipment from failing when electricity grids experience unstable electricity supply or complete supply failures. An unstable electricity supply has the potential to damage the FGIP equipment. Correctly installed UPS systems will protect the equipment against damage from most electricity supply fluctuations and outages and supply power, allowing procedures to be safely completed. The supplier of the FGIP equipment, together with the facility electrical engineer, needs to be consulted to determine the suitable size of the UPS and the required battery backup capacity. It is important that UPS systems be tested frequently and serviced routinely according to the UPS supplier schedules to ensure the systems function as expected.

3.20. ADDITIONAL TOOLS

Some additional optional tools that may be a part of the FGIP system include automatic reduction of detector-to-patient distance (a function that reduces this distance without manual intervention of the operator), virtual wedge filter positioning, virtual collimation and virtual centring tools. These features allow virtual positioning of the wedge filters, collimation, or centring of the patient without further use of X rays. Figure 23 shows an example of virtual filter positioning and virtual collimation function.

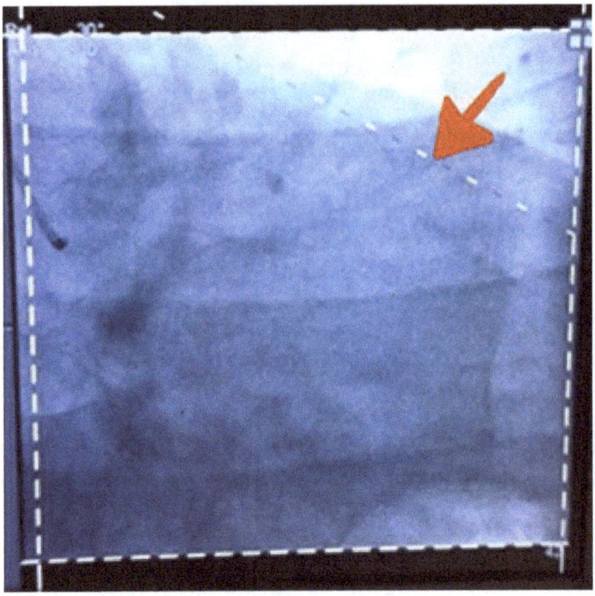

FIG. 23. Virtual filter positioning and virtual collimation.

4. QUALITY ASSURANCE PROGRAMMES

QAP is a comprehensive concept that incorporates distinct yet interconnected components of QA and QC. Each of these components plays a crucial role in ensuring the highest standards of quality and is elaborately defined in the sections that follow.

4.1. QUALITY ASSURANCE

QA is a concept that encompasses all of the factors that affect an expected result, namely a clinical diagnosis. The QA is part of the quality management system that radiological medical facilities establish. This is recognized by the World Health Organization (WHO) [31], who defined QA as "those planned and systematic actions necessary to provide adequate confidence that a structure, system or component will perform satisfactorily in service". According to the IEC, the QA implies "the optimum quality of the entire process, i.e. the consistent production of adequate diagnostic information with minimum exposure of both patient and personnel". The details of the instructions or actions that QA implies is defined as a QAP, and this is focused on providing security on that the quality standards that are defined are met [32].

Various IAEA guidelines underline the importance of a QAP [33–37] and mention that QAPs in radiological medical facilities need to include the following [37]:

(a) Measurements of the physical parameters of medical radiological equipment made by or under the supervision of a CQMP:
  (i) Before the equipment is approved for patient use and before initial set-up;
  (ii) On a regular basis following that;
  (iii) Following any significant maintenance work that might impact patient safety and protection;
  (iv) After the installation of any new software or alternations to any existing software that could impact patient safety, clinical outcome, etc.
(b) Taking corrective measures when the physical parameters identified in (a) exceed the set tolerance levels.
(c) Ensuring the physical factors applied in radiology practices are correct.
(d) Documenting relevant processes and outcomes.
(e) Conducting regular assessments of the dosimetry and monitoring equipment's calibration and operational status.

4.2. QUALITY CONTROL

QC is the part of the QAP that deals with techniques used in monitoring and maintaining the technical elements of the systems that affect the quality of image. Therefore, QC is a part of the QA and eventually part of the overall QAP. A QC programme includes the following types of performance testing [32]:

(a) Acceptance: This stage includes confirming that the equipment's specifications and functionalities meet the agreed standards and is conducted by the installer's and the facility's CQMP representatives. This process can often be as straightforward as going through a checklist. Any major deviations should be formally reported to the contractor for necessary corrective measures. Additionally, a qualified individual should evaluate the electrical and mechanical safety of the new installation during the acceptance tests.
(b) Commissioning: Performed by a facility's representative, often a CQMP with expertise in radiology physics, this process verifies the equipment's

readiness for patient care and sets initial benchmarks for future routine performance evaluations. Comprehensive testing should cover all expected clinical usage parameters and conditions. It is necessary to confirm that all pertinent evaluations of the system have been executed without omissions throughout the critical examination or acceptance phases. Following significant upgrades or repairs, such as replacing an X ray tube or installing new software, it may be necessary to conduct commissioning tests again to establish updated baseline values.

(c) Constancy: Also known as routine performance tests, these are essentially a selection of the commissioning tests and typically involve personnel with varying expertise. The simpler, quicker tests that are conducted more frequently are usually undertaken locally with advice and oversight from a CQMP. The more intricate and time-intensive tests might necessitate specialized skills and equipment.

It is essential to acknowledge that each scientific measurement and calculation inherently carries a degree of uncertainty. This factor needs to be duly considered when presenting a QC result. Specific details about uncertainties are provided in Appendix II.

4.3. ROLES AND RESPONSIBILITIES

This section of the publication is dedicated to detailing the responsibilities of key role players involved in the QAP for FGIPs. It outlines the specific duties and contributions of various professionals. Each role is critically examined to illustrate how their individual responsibilities intersect to ensure the highest standards of quality and safety, efficiency and effectiveness in FGIPs.

4.3.1. The licensee or registrant

The use of FGIP equipment, as with any X ray equipment, ought to be controlled by a regulating or licensing authority [34]. It is the responsibility of the regulating authority to issue licenses or permissions to the licensee of the FGIP facility or equipment. Depending on the requirements in each country, the licensee may be a hospital, or a business, a medical practise or a medical professional. The licensee or registrant assumes overall responsibility to ensure that the QAP is implemented and compliant for the FGIP facility or equipment and may appoint responsible persons for each QAP function. Registrants and licensees need also to ensure that regular and independent audits are made of the programme of QA for medical exposures and that their frequency is in accordance with the complexity of the radiological procedure being performed and the associated risks.

The regulator ought to require documented compliance with any legislative or regulatory provisions, established requirements and/or minimum standards. The responsibility of the regulator extends to the performance of routine inspections to provide reassurance that the requirements are fulfilled. The detailed responsibilities of the regulating authority are explained in the IAEA Safety Standards Series [34, 36].

4.3.2. Radiation protection officer

The radiation protection officer is responsible for all radiation protection and safety matters with respect to occupational and public radiation protection related responsibilities in the facility. This person needs to be a professional with sufficient training in radiation protection, such as a health physicist, CQMP, experienced MRT or any other professional with adequate radiation protection training. The occupational dose levels recorded in the FGIP environment are generally higher than other radiology areas of the hospitals and therefore requires additional and specific attention. The radiation protection officer needs to oversee the implementation of the specific radiation safety measures and needs to be knowledgeable on regulations and requirements in this regard to ensure compliance.

4.3.3. FGIP specialist (radiological medical practitioner)

As FGIP techniques develop, there is an ever-increasing variety of different specialist radiological medical practitioners utilizing FGIP equipment. It is important that these specialists are adequately trained and acquire the clinical skills necessary to perform the specialized procedures before attempting cases. The radiological medical practitioners ideally should be trained on image interpretation, as clinical decisions will often need quick decision making and intervention. The radiological medical practitioners need to be trained in the 'as low as reasonably achievable' principle, understand the displayed radiation dose indices, the risks involved (e.g. clinical, radiation, biological exposure, etc.) and have the radiation protection knowledge and skills needed to work safely in the environment [3]. The radiological medical practitioners need to be trained on the FGIP equipment being used and need to work closely and communicate clearly with the clinical team regarding the clinical and imaging requirements. The radiological medical practitioner is ultimately responsible for the safety of the patient and staff as well as the efficacy of the clinical procedure.

4.3.4. MRT

The primary role of an MRT (in some places referred to as radiological technologist, radiographer, X ray technologist) in the FGIP facility is focused on the FGIP equipment operation, and this professional ought to be adequately trained in FGIP equipment operation and radiation protection for patient and occupational safety. The MRT needs to understand the clinical goals for each procedure and needs to communicate to the radiological medical practitioner any limitations in the equipment operation or radiographic imaging technique applied.

The MRTs ideally should be able to perform QC procedures within their scope of profession and can be responsible for the QC tests described in Section 5.1 of this publication under the supervision of CQMP. These tests are generally performed before routine clinical operation. They are frequent tests that are quick to perform and do not require advanced testing equipment, or in-depth analysis or calculations.

4.3.5. CQMP

The CQMP focuses on the technical aspects of the QAP and is responsible for to developing and implementing programmes to ensure that the FGIP system meets its technical specifications and minimum requirements before being accepted for clinical use. The CQMP is also responsible for routine performance testing to ensure continued compliance of the FGIP equipment [37]. The CQMP is responsible for the advanced QC tests described in Section 5.2 of this publication.

The CQMP needs to be involved during the different phases of the life cycle of FGIP equipment, which include evaluating procurement options, overseeing facility building and equipment installation, performing acceptance testing, commissioning and routine performance testing. It is especially important for the CQMP working in FGIPs to be involved during the clinical FGIP operation and proactively lead QAP goals together with the radiological medical practitioners, MRTs and other clinical disciplines that may utilize the FGIP facility [38]. In the case of FGIPs, it is very important for the CQMP to have an adequate understanding of the clinical aspects of procedures performed using the fluoroscopy systems under supervision.

4.3.6. Service engineer

The service engineers, whether clinical engineers or original equipment manufacturer engineers, have a significant role in ensuring that the equipment is properly maintained. Engineers ought to be appropriately trained to support all the technical aspects of the equipment for which they are responsible and, ideally, should follow at least the original equipment manufacturer schedules for routine maintenance.

4.3.7. Ancillary personnel

The large variety of procedures being offered in FGIP facilities involves many different professionals who may require staff rotation at times. These additional personnel include nurses, nursing assistants, technologists and supplier representatives, among others. Although additional personnel required in the FGIP environment may not have specific roles and responsibilities related to the QAP goals, they ought to be properly educated and trained and ought to have a good understanding of radiation risks and radiation protection requirements in the facility.

4.4. TRAINING REQUIREMENTS

There are three primary categories of trainees for FGIPs:

(a) FGIP equipment operators[1]: those individuals who operate or direct the operation of the FGIP equipment during clinical use. Radiological medical practitioners who instruct an MRT regarding what imaging they want are considered operators for the purpose of this training.
(b) Non-operator radiation workers: those individuals who are in the procedure room and occupationally exposed to radiation but do not operate the FGIP equipment. Examples of this type of trainee are anaesthesiologists and nurses who are present in the FGIP suite during the procedure but do not operate or adjust the fluoroscopy equipment in any manner.
(c) Non-radiation workers with access: those individuals who are not considered radiation workers and are not present during the FGIP procedure but may enter the fluoroscopy suites at other times. Examples of this type of trainees include environmental services (e.g. cleaning staff), clinical engineering (i.e. equipment service technicians) and facilities management (e.g. building maintenance staff).

[1] This publication addresses only operators who perform FGIPs. Differentiated, reduced levels of operator training would be appropriate for individuals who only perform routine diagnostic fluoroscopy procedures, or those individuals who only perform extremity imaging using mini C-arm fluoroscopy equipment.

Training content needs to be commensurate with the clinical need. Therefore, training for category 1 (as per Table 1 below) will be substantially greater than categories 2 and 3 which ought to be limited to information pertinent to those individuals' safety. Table 1 lists the suggested training topics for different categories of trainees.

In addition to the general training identified above, category 1 trainees need to have machine specific training (see Table 2) for each make and model of fluoroscopy equipment that they operate. This training is intended to ensure that operators are familiar with the system controls and functions, often referred to as 'buttonology'. These controls are not consistent between vendors and sometimes not even consistent between models of equipment for the same vendor.

TABLE 1. TRAINING TOPICS FOR DIFFERENT CATEGORIES OF TRAINEES

No.	Topics	Category
1	Basic principles of radiation protection and optimization	1
2	Radiation quantities and units pertinent to fluoroscopic imaging and dosimetry	1
3	Radiobiology pertinent to fluoroscopy	1
4	Brief overview of basic fluoroscopic imaging technologies	1, 2*
5	General operator controlled factors: common features	1
6	Radiation indicators and information displays	1, 2, 3*
7	Paediatric patient considerations and concerns	1
8	Pregnant patient considerations and concerns	1
9	Pregnant workers and occupational exposure	1, 2
10	Management of individual patient radiation dose during procedures	1
11	Post-procedure facility processes for the management of individual patients with the possibility of tissue reactions	1, 2
12	Documentation of patient radiation parameters	1

TABLE 1. TRAINING TOPICS FOR DIFFERENT CATEGORIES OF TRAINEES (cont.)

No.	Topics	Category
13	Occupational radiation dosimetry	1, 2
14	Occupational radiation protection	1, 2
15	Radiation fields tableside	1, 2
16	Radiation sources and signage (hazard identification)	1, 2, 3
17	Local/regional/national regulations directly affecting fluoroscopy personnel	1, 2*
18	Facility policies and procedures for pre-procedure patient education, intra procedure dose management, post-procedure patient education and follow-up requirements	1

Note: * indicates limited or reduced content depth needed for that category.

TABLE 2. MACHINE SPECIFIC TRAINING FOR CATEGORY 1 TRAINEES

No.	Topics
1	Exam protocol selection
2	All switches and controls that initiate or control X rays (e.g. foot switches, hand switches, etc.)
3	All functions that affect the imaging or radiation output (e.g. magnification, low dose fluoro, high dose fluoro, collimation, wedges, pulse rates, etc.)
4	Mechanical movements/controls (e.g. table movement, gantry movement, SID changes, etc.)
5	Emergency stop, overrides, breakers (only test emergency stop if service is readily available)
6	Radiation disable function

TABLE 2. MACHINE SPECIFIC TRAINING FOR CATEGORY 1 TRAINEES (cont.)

No.	Topics
7	Grid removal/replacement
8	Monitor/display layout if changeable
9	‘Fluoro Save’ and ‘Last Image Hold Save’ operation
10	Displayed radiation dose indices and available units of measurement

4.5. EQUIPMENT CYCLE MANAGEMENT

QA for FGIP equipment, as with any imaging equipment, does not end with annual or regulatory compliance testing by a CQMP. Several reports have been published on this topic, and readers are encouraged to review those sources for additional in-depth information [39–43]. The primary need is for an overarching QAP establishing roles for overseeing all aspects of the equipment life cycle and clinical use. Ideally, a QAP will define the following roles:

(a) Physician (radiological medical practitioner) champion/leader.
(b) CQMP.
(c) MRT or expert user.
(d) Clinical engineer/service engineer.
(e) Department administrator.
(f) Ancillary members to be included on an as needed basis, but need to understand the specific needs and challenges pertaining to FGIPs:
    (i) Radiation protection officer or health physicist;
    (ii) Informatics specialist;
    (iii) Construction management;
    (iv) Equipment vendor specialist;
    (v) Consumer (optional, but good to periodically have a patient’s perspective on different points in the process).

Although discussion of a full QAP is largely outside of the scope of this publication and warrants its own report, it is applicable because it is central to the equipment life cycle and management of that equipment.

The core membership of the QAP [19, 34, 35, 39–43] need to be engaged during all stages of the equipment lifecycle as each brings different perspectives and training. Figure 24 is a diagram adapted from several sources [41–43]. Descriptions of the diagram are as follows:

(a) At the beginning of the cycle is the exploration and evaluation of the current state of the technology and how that fits with the anticipated or projected clinical use and workflow. Considerations at this point include the quality of the imaging system, service availability (ensuring that the equipment vendor adequately support the system), advanced imaging tools that may be needed or desired, informatics integration, and regulatory compliance or applicable standards (does the system meet local, regional, national and/or international standards for the types of clinical procedures to be performed). As an example, since this publication pertains specifically to FGIPs, all FGIP equipment ideally should meet the IEC 60601-2-43 standard, which has specific requirements for FGIP equipment intended for use in these types of procedure [9]. Other such standards or regulations may apply and ought to be reviewed during this stage of the equipment cycle.

(b) The procurement phase of the cycle involves ensuring that the appropriate imaging system(s) are included, with the necessary tools for clinical efficacy, patient and personnel safety, and regulatory compliance. Additionally, considerations are needed at this phase for necessary service contracts (including service level measurables like key performance indicators and machine uptime guarantees) and vendor-provided training to ensure safe and effective use of the equipment.

(c) The acceptance testing phase is traditionally the phase in which the CQMP performs an extensive set of tests to ensure that the imaging equipment functions as designed and intended and that it meets all applicable regulatory requirements or standards. However, it is important that all members of the QAP team evaluate the systems and structures to ensure safe and effective use of the systems, efficient workflow and interoperability of the systems with each other and within the larger hospital framework.

(d) The commissioning phase is essentially preparing the now accepted system(s) for clinical use. As a starting point, baseline values are defined. After this stage, the system needs to be prepared for clinical use. This process requires input from all clinical personnel and vendor specialists to ensure that optimized protocols are used for the procedures performed and personal preferences of the staff. This optimization ideally should consider all fluoroscopy equipment protocol technical factors (e.g. pulse rates, default fluoroscopy curves, acquisition rates, necessary image quality and dose), image processing, display layouts, table and foot control configurations, room lighting, operator shielding and placement, etc.

(e) Once commissioned, ongoing analysis and feedback to the QAP are essential for all systems. Clinical use presents many new variables and challenges that commissioning may not have been able to address, requiring changes to the protocols or system settings to maximize safety and efficacy.

(f) All FGIP systems need to be delivered with numerous documents, one of which needs to describe the vendor maintenance schedule. It is essential that responsible organizations perform or ensure that the routine maintenance of these systems are performed for the safe and effective use of the systems. Whenever substantial service is performed, especially replacement of major components (e.g. X ray tube, image receptor, etc.), re-evaluation occurs and potentially adjustments to the protocols, as needed. This part would also pertain to routine QA performed by appropriately CQMP and/or MRTs as appropriate.

(g) Disposal of the systems, though seemingly straightforward, has its own set of concerns from the perspective of a QAP. First, it ought to be ensured that all necessary information has been transferred off the system and that all information that can be removed is removed, especially patient information. Secondly, the equipment needs to be responsibly disposed of. and it must be ensured that it does not present a hazard to anyone involved in the process or ultimately to the public. Again, at this final stage, there may be local, regional, national or international regulations that need to be followed, and it is important for all team members to understand what is required.

5. PERFORMANCE TESTS

The MRTs can perform certain QC tests within their scope of profession under the supervision of CQMP. These tests are typically conducted prior to regular clinical operations. They are routine and quick to execute and they do not necessitate sophisticated testing apparatus or detailed analysis and calculations.

5.1. MRT TESTS

All persons performing these tests need to be adequately trained on the proper techniques of performing the tests. They also need to be aware of all

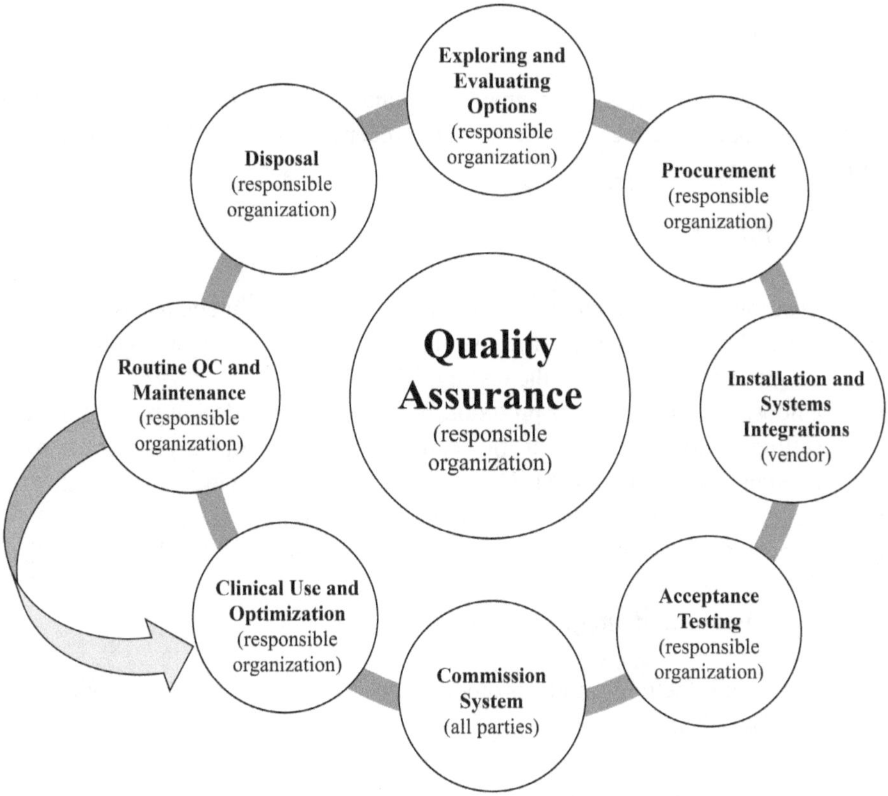

FIG. 24. Equipment cycle management. This diagram is adapted from multiple sources [39–43].

applicable personal radiation safety (i.e. use of lead or lead-free protective equipment and auxiliary shields).

5.1.1. Reproducibility of AERC

5.1.1.1. Description and objective

AERC is mandatory for all fluoroscopy systems. It automatically adjusts the exposure parameters (i.e. kVp, mA, pulse length, filtration) to obtain a constant predefined image quality [44].

5.1.1.2. *Equipment*

Twenty centimetres of polymethyl methacrylate (PMMA) slabs or copper plates positioned to cover the whole image receptor area or collimate to the size of the attenuator.

5.1.1.3. *Procedure*

(a) Position the chosen attenuator on the table or on the image receptor.
(b) Note the exact set-up (focus-to-image receptor distance, table height, longitudinal position for the testing, FOV, etc.)
(c) Choose a clinically used protocol, make sure the grid is in place, note the settings (mode, dose level, pulse frequency, etc.) and expose for a long enough time so that the AERC reaches a stable state with constant values.
(d) Record $K_{a,r}$ rate (mGy/min), kVp, mA, filtration.
(e) Repeat with two other thicknesses of the attenuator and for different FOVs.

5.1.1.4. *Analysis and interpretation*

For each setting and protocol, compare the readings with the established baseline.

5.1.1.5. *Baselines and tolerances*

(a) The tolerance with respect to the baseline is ±5% for the kVp and ±20% for the mA.
(b) The other parameters ought to be the same.

5.1.1.6. *Frequency*

Quarterly (every 3 months).

5.1.1.7. *Corrective actions*

Ask the CQMP to perform a more thorough investigation if the deviation persists, as this could severely impact image quality. Ultimately, if the error is reproducible, the service engineer ought to be contacted. The X ray machine ideally should not be operated without an appropriately working AERC unit.

5.1.2. Artefact evaluation

5.1.2.1. Description and objective

Image artefacts assessment is a process of evaluating imaging studies to identify any artefacts that may have been introduced during the imaging process or any flaws in the image caused by certain conditions of the environment or the scanning device. This assessment is used to determine the accuracy and reliability of the image and can help identify factors that may contribute to a misdiagnosis or incorrect conclusions from an image. The assessment can also help ensure that radiological medical practitioners and other observers accurately read and interpret the image in order to make an accurate diagnosis.

5.1.2.2. Equipment

Flat field images are acquired using PMMA or other uniform medium to stimulate typical system settings, corresponding to the technical parameters employed for an average patient size, used within the clinical range and providing an image that is as uniform as possible as an input.

5.1.2.3. Procedure

Review the images carefully for the presence of any artefacts (dirt, blood, contrast agent on image receptor or other surface in the X ray beam path, table or monitor, unexpected object overlaid on anatomy).

5.1.2.4. Analysis and interpretation

Visible avoidable artifacts need to be absent. An example of an artefact is shown in Fig. 25.

5.1.2.5. Baseline and tolerances

Not applicable.

5.1.2.6. Frequency

Quarterly (every 3 months).

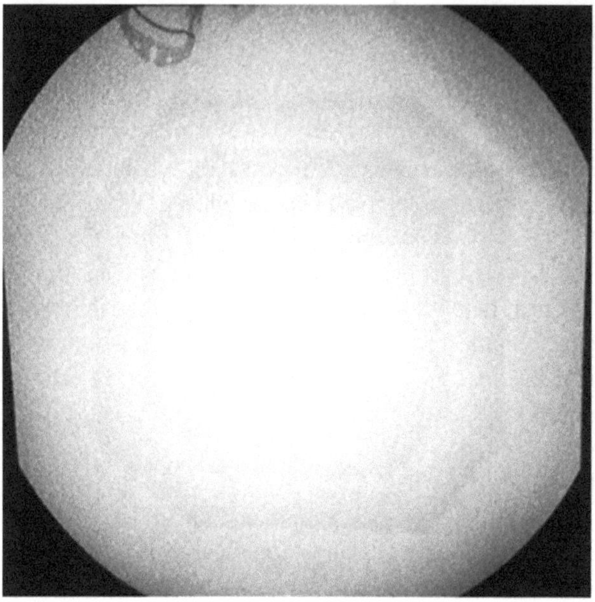

FIG. 25. Example of an artefact: smaller FOVs are visible on the image.

5.1.2.7. Corrective actions

Contact the CQMP to evaluate the severity of the artefact and decide the follow up actions. If a phantom has been used, it is useful to rotate the phantom to ensure that the artefact is not inherent. Alternately, the interventionalist can be consulted to assess clinical impact before continuing clinical use.

5.1.3. Monitor display check

5.1.3.1. Description and objective

The display monitors are an important part of the imaging chain in any FGIP system, and this test is designed to be a quick qualitative evaluation of the performance of the display system. It provides reassurance to the MRT that the display system is functioning as expected with no obvious faults exceeding the action levels.

5.1.3.2. Equipment

This test requires the display of a suitable test pattern. The following test patterns are suggested for this test: AAPM TG18 QC [15], IEC modified TG18-OIQ [45] and AAPM TG270 sQC [46].

The older versions of the Society of Motion Picture and Television Engineers medical diagnostic imaging test pattern are not endorsed by this publication for the purposes of testing LED monitors, as they were designed for older CRT monitors that are no longer manufactured and not covered in this publication.

5.1.3.3. Procedure

Load the selected test pattern on all display monitors used clinically during routine operation of the FGIP equipment. Evaluate the qualitative features of the chosen test pattern. The procedure of evaluation will vary depending on the type of test pattern being used, as they have different features.

Very large displays with configurable layouts are common in FGIPs. For these displays, the user can discuss the expected resolution levels with the supplier and evaluate the most used clinical layout, as evaluating all possible layouts are not feasible [46]. The following procedures can be applied while using different test patterns:

(a) For TG18-QC [47] and TG18-OIQ test patterns [45]:
   (i) Perform a qualitative evaluation of the low contrast sections to evaluate the luminance response function. Pay attention to the following sections:
      — 0%/5% and 95%/100% contrast objects.
      — 'QUALITY CONTROL' letters visible at different levels of background contrast.
      — Sixteen luminance patches and their smaller contrast corner patches.
      — Luminance ramps along the sides of the patterns.
   (ii) Inspect the displayed pattern for artefacts or abnormalities.
(b) For TG270-sQC test pattern [46]:
   (i) Visually assess the luminance response and pay attention to:
      — Three rows of greyscale patches (total 18 patches) with incrementing luminance.
      — Evaluate the bar pattens in the upper left and lower right corner of each greyscale patch. These bar patterns are +/-5 grey levels different from the background.

— The greyscale gradient at the bottom of the pattern.

(ii) Inspect the displayed pattern for artefacts or abnormalities.

5.1.3.4. Analysis and interpretation

The analysis and interpretation of this test is qualitative in nature and therefore relies on and assumes the normal functioning of the human visual system of the MRT performing the evaluation. Nonetheless this qualitative test is a valuable tool to evaluate the correct functioning of the display monitors because it is relatively quick to perform.

5.1.3.5. Baselines and tolerances

The baselines and tolerances are discussed for each pattern below:

(a) TG18-QC and TG18-OIQ test patterns:
   - (i) 5% and 95% inserts ideally should be visible in normal light settings.
   - (ii) 'QUALITY CONTROL' letters ought to be visible at the different levels of background luminance.
   - (iii) All 16 luminance patches need to be visible.
   - (iv) No artefacts compromising clinical use ought to be present on the display.

(b) TG270-sQC test pattern:
   - (i) All 18 greyscale patches ought to be distinguishable.
   - (ii) All the +/-5 grey level bar patterns inside of the 18 greyscale patches need to be visible.
   - (iii) No artefacts compromising clinical use ideally should be present on the display.

5.1.3.6. Frequency

Quarterly (every 3 months).

5.1.3.7. Corrective actions

Contact the CQMP to evaluate the severity of the display malfunction and, if necessary, perform an advanced evaluation of the display as described in Section 5.2.12. Facilities with multiple displays may consider restricting the clinical use of the affected display so that the FGIP systems can still be used while the CQMP evaluation or the repair or replacement is underway.

5.1.3.8. *Clinical considerations*

The incorrect functioning of the image display monitors of an FGIP system, outside of the tolerance levels, can impact the user's ability to visualise and interpret the images correctly during procedures. This may lead to missed diagnoses or even mistakes during interventions that can negatively affect the clinical outcome of patients.

5.2. CQMP TESTS

All persons performing these tests ideally should be adequately trained on the proper techniques of performing the tests as well as all applicable personal radiation safety (i.e. use of lead or lead-free protective equipment and/or auxiliary shields). Concerning the timing of the tests detailed below, the majority are necessary following significant maintenance or servicing events.

5.2.1. System checks

5.2.1.1. *Description and objective*

The objective of the systems checks is to ensure proper operation of the mechanical and functional components of the FGIP system.

5.2.1.2. *Equipment*

No equipment is necessary for the performance of this evaluation.

5.2.1.3. *Procedure*

Activate all system controls, including all mechanical and electronic controls, pressure sensors on the system that prevent system movement for patient and operator safety, display controls, illuminated signs or X ray beam indicators that illuminate or otherwise indicate when the X ray beam is active, X ray disable, system alarms, etc. Although a necessary control to evaluate, caution needs to be used when evaluating the 'emergency off' switches, it is best to check these controls only when service support is immediately available. These switches may malfunction and not allow the system to return to normal functionality without service repair.

5.2.1.4. Analysis and interpretation

FGIP equipment needs to be capable of all intended functionality. The system ideally should be fully intact, including the housing for all components, as the housing for the C-arm and imaging chain components protects the electronics for these components from potentially infiltrating liquids or debris.

5.2.1.5. Baselines and tolerances

Any non-functional or compromised controls or components need be repaired to bring the system to its intended state.

5.2.1.6. Frequency

Acceptance testing ought to involve extensively evaluating all system controls and components that are available, within reason (the evaluator does not require tools or service level access to perform evaluations as part of this test).

5.2.1.7. Corrective actions

Service ought to be requested for any identified issues.

5.2.2. X ray beam alignment and centring

5.2.2.1. Description and objective

The aim of X ray beam alignment and centring is to ensure that the X ray field is correctly sized and precisely aligned with the image receptor. Proper alignment of the central ray with the centre of the image receptor minimizes image cutoff and avoids unnecessary exposure to surrounding tissues. Accurate X ray beam alignment and centring not only prevent unnecessary exposure to patients but also reduce scattered radiation to medical personnel while preserving or even improving image quality.

5.2.2.2. Equipment

— Collimation/alignment test tool;
— Measuring tape.

5.2.2.3. *Procedure*

(a) If the C-arm is capable of completely inverting, allowing the X ray tube to be positioned above (possible for most mobile systems, but few fixed C-arms), this orientation is preferable. Otherwise, position the C-arm in a vertical position (0 degree).

(b) Select the largest possible field size and fully extend the collimators to their maximum opening.

(c) If the C-arm can be inverted, then place the plate directly on the image receptor. Otherwise, place the test tool on the procedure table, with the plate as close to the image receptor as possible ensuring it is placed on top of the cylinder and remains level (see Fig. 26). If the test tool is not placed directly on the image receptor, measure the distance between the test tool and the focal spot, then calculate the magnification factor. (Using an object of known size, such as a coin, can allow determination of magnification, if distance cannot be measured). Also ensure that the procedure table and test tool are level.

$$\text{Magnification factor} = \frac{\text{Focal spot - } t_0 \text{- image distance}}{\text{Focal spot - } t_0 \text{- object distance}} \tag{6}$$

(d) Using fluoroscopic guidance, adjust the procedure table until the points within the cylinder are aligned and overlapping. This alignment represents the centre of the X ray beam (see Fig. 27).

(e) Measure the distance between the central ray and the centre of the image receptor in both the lateral and vertical directions.

5.2.2.4. *Analysis and interpretation*

Determine the distance between the central ray and the centre of the image receptor in both the horizontal and vertical directions.

5.2.2.5. *Baselines and tolerances*

The deviation of the X ray field from the image receptor area is not to exceed 2% of the SID.

5.2.2.6. *Frequency*

Annually.

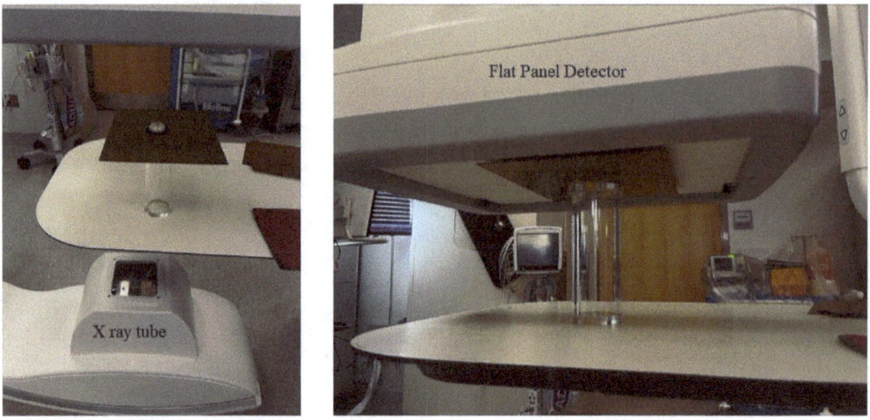

FIG. 26. Positioning of the X ray beam alignment and centring test tool.

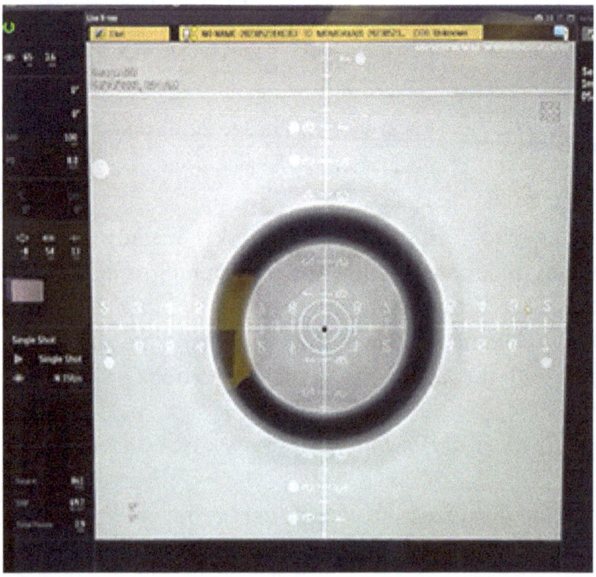

FIG. 27. Alignment of the central points within the cylinder.

5.2.2.7. *Corrective actions*

Check the primary and secondary gantry angles to ensure that the system is at 0 degrees in both rotational directions.

5.2.3. X ray beam collimation

5.2.3.1. Description and objective

This test ensures that the X ray field is correctly confined and does not extend beyond the boundaries of the image receptor assembly or the designated viewable area.

5.2.3.2. Equipment

- — Large lead plate.
- — Radiochromic film. If radiochromic film is not available, then a CR cassette or fluoro screen can be used.
- — Radiopaque ruler.
- — Tape.

5.2.3.3. Procedure

(a) Ideally, the X ray beam collimator blades will be adjusted such that they are visible, or their penumbra is visible, in each FOV. **If this criterion is met, then nothing further needs to be done, and this test is automatically passed.**

(b) Rotate the C-arm into the lateral position (or invert the unit if this is possible).

(c) Although this test can be performed at any SID, a SID of 100 cm is preferred.

(d) Use a radiopaque ruler to determine the X ray field sizes at the face of the image receptor assembly and record the values in both directions (see Fig. 28).

(e) Determine a magnification factor using the lead ruler or other object of known size.

(f) With the C-arm in the lateral position, move the procedure table as close to the image receptor as possible.

(g) Support the lead plate on the procedure table and tape the lead plate on the image receptor, ensuring that the lead fully covers the image receptor at the largest FOV (Fig. 29(a)).

(h) Once the lead is properly placed, add radiochromic film to the set-up (Fig. 29(b)).

(i) Begin with the smallest FOV and irradiate the film in acquisition mode either cine or DSA.

(j) Repeat step (i) for each available FOV. Example of an exposed radiochromic film with multiple tested FOV is shown in Fig. 30.

5.2.3.4. Analysis and interpretation

On the film, measure the field sizes and compare to the nominal field sizes. (Note: some vendors report/display the diagonal dimension of the image receptor, while others show the length of one side of the square.)

5.2.3.5. Baselines and tolerances

The sum of deviations ought not to exceed 2% of the SID in either direction.

5.2.3.6. Frequency

Annually.

5.2.3.7. Corrective actions

If the tolerances are exceeded, initiate engineering service.

5.2.4. kVp accuracy

5.2.4.1. Description and objective

In FGIPs, the tube voltage is one of the most important parameters affecting both radiation doses and diagnostic image quality for the patient. The objective of this test is to evaluate the accuracy of the tube voltage selected from the operator's console of the FGIP system [43].

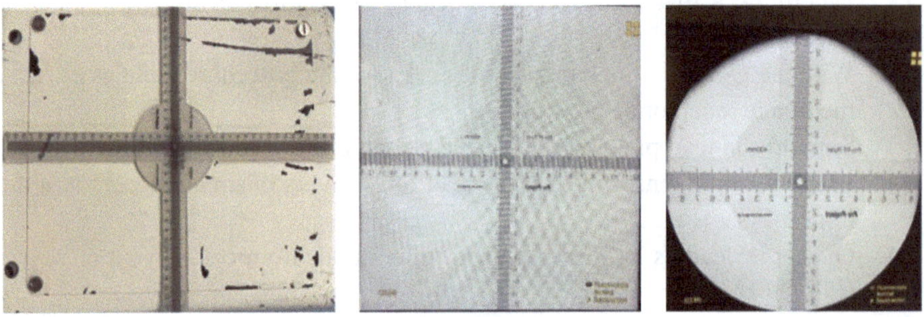

(a) (b) (c)

FIG. 28. (a) Radiopaque ruler placed in front of the image receptor assembly for X ray field size measurement, (b) image obtained from an FPD and (c) image obtained from an image intensifier.

(a) (b) (c)

FIG. 29. (a) A lead plate is placed in front of the image receptor. (b) Radiochromic film is placed on the lead plate. (c) Alternatively, the radiochromic film may also be cut into four quadrants and taped on the four quadrants of the image receptor (ensuring that the radiochromic film coverage exceeds the largest FOV size).

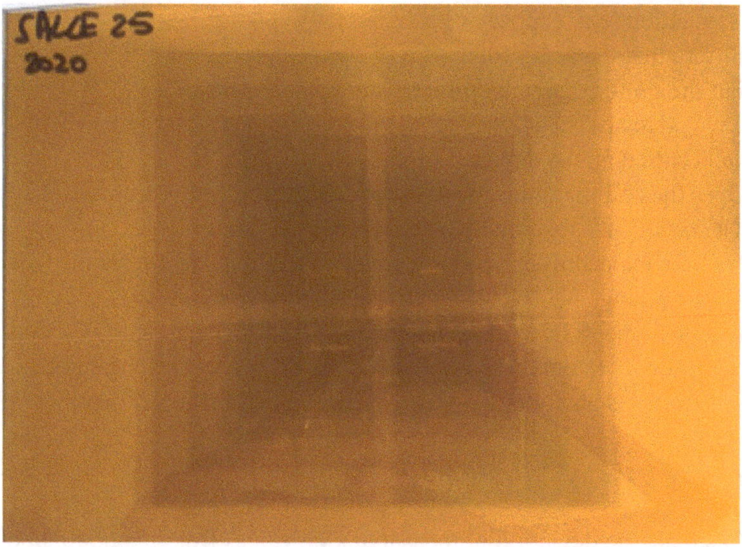

FIG. 30. Example of an exposed radiochromic film with multiple tested FOVs.

5.2.4.2. *Equipment*

An appropriate X ray tube voltage measurement device, which has a specified range (e.g. 40–150 kVp) that covers the full range of kVps of which the system is capable. Most kVp measuring devices have different options for deriving a kVp, such as the practical peak voltage, the average kVp or maximum kVp. Either a solid state kV meter or solid state multimeter with accuracy ±1 kV and reproducibility ±1% needs to be used. Instruments based on non-invasive measurements are preferred for use. The use of attenuators (such as copper or lead plates) to protect the image receptor or to replicate different levels of patient thickness is also suggested [3, 32, 44].

5.2.4.3. *Procedure*

(a) If available (preferable), use a service mode or manual radiographic mode to manually select the technique parameters. Otherwise, this test can be done by using various amounts of copper or PMMA attenuation on the image receptor side of the kV meter to avoid hardening the measured X ray beam.
(b) Set and register 4 or more clinically relevant kVp values across the full available range (e.g. 60, 80, 90, 110 kVp) as well as a relatively low mA value (e.g. 2 mA).
(c) To avoid damage, cover the image receptor with an appropriate amount of attenuation (e.g. copper or lead plate).
(d) Ensure that the X ray tube is positioned correctly.
(e) Put the non-invasive kVp meter on a stable surface with its sensitive area directed towards the X ray tube. Ensure that the X ray beam is perpendicular and centred on the device.
(f) Open the collimator in a way that covers the sensitive volume of the detector. Then check its position using a fluoroscopic image.
(g) Perform exposure for every chosen kVp and register the results (see Fig. 31 for the arrangement).
(h) If manual mode selection is not possible, use attenuators, FOVs, dose curves or protocols to obtain the desired kVp and mA values. Make sure that the kVp meter is positioned correctly. See Fig. 32 for the arrangement [3, 44].

5.2.4.4. *Analysis and interpretation*

Calculate the deviation of measured kVp values from the nominal ones.

5.2.4.5. *Baselines and tolerances*

Maximum deviation ±5% or 5 kVp, whichever is greater.

5.2.4.6. *Frequency*

During acceptance testing, annually, and after each change of the X ray tube.

5.2.4.7. *Corrective actions*

If the tolerances are exceeded, first repeat the test. If the problem persists after repeating the test, contact the supplier of the X ray system.

5.2.5. mAs linearity

5.2.5.1. *Description and objective*

Radiation output is defined as the ratio of the air kerma to the product of the tube current and time (mGy/mAs) at a fixed X ray beam spectrum (kVp, filtration) and fixed distance. This ratio needs to remain constant over a range of current–time product values.

This test serves as an indirect method for assessing the accuracy of the X ray tube current. Direct measurement of X ray tube current involves invasive

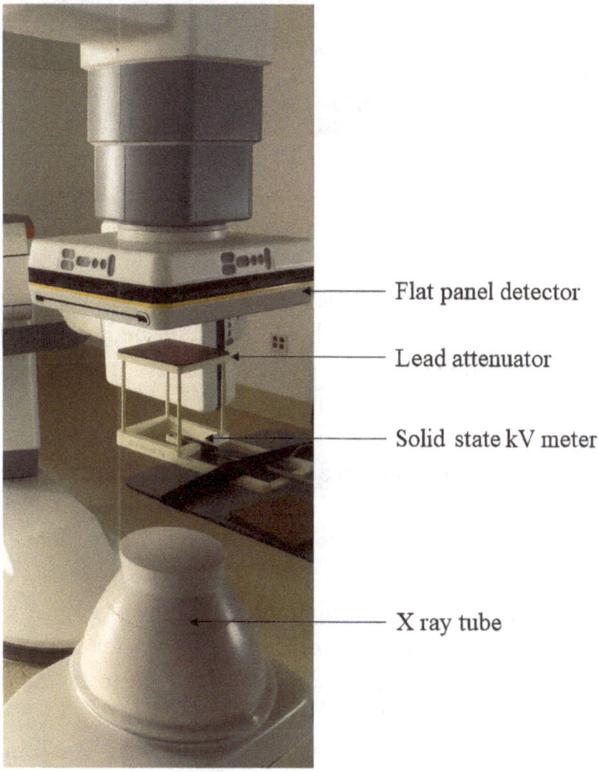

FIG. 31. Test set-up for kVp accuracy using a solid state non-invasive kV meter with manual selection mode of kV.

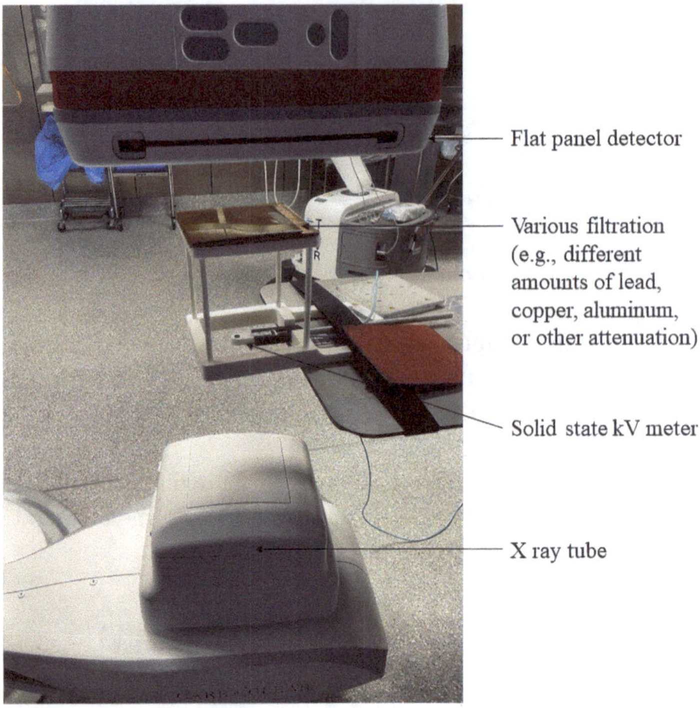

FIG. 32. Test set-up for kV accuracy using a solid state non-invasive kV meter with the automatic control system.

procedures that require access to the high voltage generator power supply, which is highly dangerous and should be performed only by qualified service engineers. The mAs linearity test, however, offers a safe alternative by demonstrating that changes in the set mAs result in a relatively consistent mGy/mAs. This consistency indicates that the tube current is adjusting as expected, allowing for accurate performance evaluation without direct measurement.

5.2.5.2. Equipment

An appropriate radiation measurement device to measure the air kerma is required. This may be a solid state multimeter or ionization chamber. Regardless of the measurement device used, it needs to be calibrated for the X ray beam spectra being measured.

X ray attenuating material in relatively small increments is required, with sufficient total filtration to cover the clinically used range. An example would be eight copper filters of 1 mm thickness, and one filter of 0.5 mm thickness. This would allow an individual to test the full range of a system in 0.5 mm Cu increments. It is not necessary to perform this test with copper; other materials are suitable. Most important is that the range of attenuation covers the equivalent range of clinical attenuation.

5.2.5.3. Procedure

(a) Centre the measuring instrument within the X ray field, free-in-air, if possible.
(b) Collimate the X ray field as small as reasonably achievable, while taking care to fully irradiate the measuring device.
(c) For this test, it is preferable to use the service mode or an imaging that allows for manual control of the generator settings (kVp, mA, s) and, in those systems with dynamic filtration, it is important that the filtration remains constant throughout the test. Adequately protect the image receptor with lead attenuator (see Fig. 33).
(d) Furthermore, ensure that the same focal spot setting is used for all measurements, otherwise the test may not yield appropriate results.
(e) For systems where independent control of the technical parameters is possible:
    (i) Position lead or another highly attenuating material within the X ray beam to completely shield the image receptor during exposures. Ensure this material is not placed too close to the measurement device; maintain a minimum distance of 10 cm between the attenuation material and the measurement device. This spacing is particularly important when using an ionization chamber for measurements.
    (ii) Set a fixed clinically used kVp (e.g. 80 kVp) and Cu filtration (if selectable).
    (iii) Set an mA and time or mAs on the low end of clinical use (e.g. 20 mA and 100 ms, or 2 mAs).
    (iv) Make an exposure at these settings, record the mAs and air kerma (mGy or μGy).
    (v) Repeat steps (b)–(d) for at least two additional higher mAs values used clinically (ensure the same focal spot setting is used).
(f) For fluoroscopic systems that cannot manually set the generator parameters, this test can be accomplished in a clinical protocol, preferably using a single shot acquisition. The clinical protocol ideally should be chosen such that the kVp remains constant over as large a range of attenuation as possible.
    (i) Place a small amount of attenuating material in the beam (e.g. 1 mm Cu), between the measuring device and the image receptor, ensuring at least 10 cm separation from the attenuator and the measurement

device. Also ensure that the attenuating material fully covers the active X ray field.

(ii) Make a single shot exposure and record the air kerma and mAs.

(iii) Place an additional incremental amount of attenuating material (e.g. additional 0.5 mm Cu) in the beam and make another exposure.

(iv) Repeat steps (b)–(c) for several increasing amounts of attenuating thickness, ensuring at each step that the beam quality (kVp and Cu filtration) is held constant.

5.2.5.4. *Analysis and interpretation*

Calculate radiation output (μGy/mAs) for all exposures.

5.2.5.5. *Baselines and tolerances*

(a) The radiation output between the two mAs settings, which differs by less than a factor of 2, needs to be linear, and the coefficient of linearity ought to be within ±10%.

(b) Deviations compared to baseline data acquired at acceptance testing on the same unit ought to be within ±20%, assuming no hardware changes have occurred (e.g. X ray tube replacement). If the radiation output has changed by ±50%, the cause needs to be determined before returning to clinical use. For this comparison, though, it is important for the X ray beam spectra to be the same in order to allow for appropriate comparisons.

5.2.5.6. *Frequency*

Acceptance test, annually thereafter, and following major repair or replacement of the X ray tube or generator.

5.2.5.7. *Corrective actions*

If any of the tolerances are not met, then service ought to be requested to find and correct the issue. Causes for failure may include problems with the X ray tube, generator or image receptor inasmuch as the image receptor signal is often used in AERC.

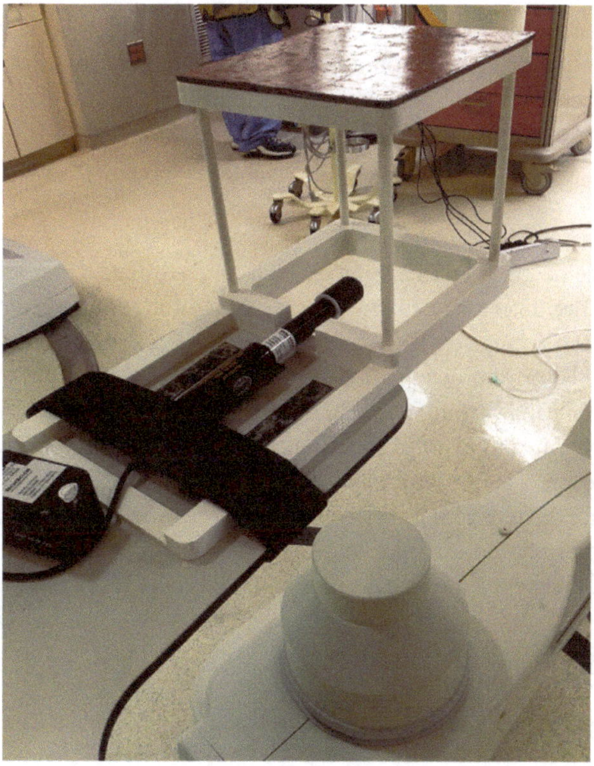

FIG. 33. Free-in-air set-up and lead protection for the image receptor.

5.2.6. Radiation output reproducibility

5.2.6.1. Description and objective

Radiation output parameters are checked to ensure their accuracy and reproducibility. kVp, exposure time and radiation output testing evaluate the consistency of the generator. The consistency of the system performance is essential to guarantee a consistent dose and image quality over time.

5.2.6.2. Equipment

kVp, total filtration, half value layer (HVL), time, air kerma rate and air kerma can be measured in a single exposure through a properly calibrated solid state multimeter.

For reasons of practicability and to save time, if no manual mode is available, these measurements are best done in service mode. In that case, the image receptor needs to be protected with a lead sheet. If service mode is not possible, attenuators ideally should be put into the path of the X ray beam behind the measurement device to achieve the desired kV value (for fixed angiographic equipment, for example).

5.2.6.3. Procedure

(a) Place the detector on the X ray table or on the image receptor. The sensitive area of the detector needs to face the X ray tube in the centre of the field. Record the settings for kV, total filtration, HVL, exposure time (one exposure is needed) for fluoroscopy and single shot exposure.
(b) Select kV values from 50 to 120 by steps of 10 (or put attenuators in the path of the beam, behind solid state multimeter), expose (fluoroscopy mode and then acquire a single shot exposure) and record the results.
(c) Calculate the reproducibility of kV, exposure time (for single shot exposure) and air kerma rate or air kerma.
(d) At 80 kV, repeat exposure 4 times and record the results.

5.2.6.4. Analysis and interpretation

(a) Calculate the deviation of measured parameter values from the nominal ones.
(b) Calculate the coefficient of variation for the parameters recorded during the four repeated exposures.
(c) Calculate output with the following equation:

$$Y = \frac{M_c}{Q} \times \left(\frac{d_{FDD}}{d_{ref}} \right)^2 \qquad (7)$$

where
Y is the radiation output (mGy/(mA·s));
M_c is the reading of the instrument (μGy);
Q is the charge or mAs;
d_{FDD} is the focal spot to instrument distance (m);
and d_{ref} is the reference distance (m; in this case, d_{ref} = 1 m).

(d) Calculate the reproducibility of kV, exposure time and output using the following equation:

$$\frac{\text{(measured value - mean value)}}{\text{mean value}} \times 100\% \quad (8)$$

(e) Check that HVL and total filtration measurements remain the same as the baseline values.

5.2.6.5. Baselines and tolerances

Measured kilovoltage values should not deviate from the nominal values by more than ±5% or ±5 kV, whichever is larger, according to established remedial guidelines. If deviations exceed these limits, corrective measures are recommended. The Institute of Physics and Engineering in Medicine (IPEM) [42] proposes a suspension level of ±10% or ±10 kV, whichever is greater and beyond which the equipment should not be used until fixed. The radiation output for 80 kV and 2.5 mm Al total filtration needs to be in the range of 25 μGy/mAs to 80 μGy/mAs. With increased filtration, outputs could drop to as low as 6 μGy/mAs. Output deviations ideally should remain within ±20% of the commissioning baseline to avoid remedial action and within ±50% to prevent suspension of use.

COV is to be less than 0.05 (5%).

5.2.6.6. Frequency

Annually and after major repair or maintenance that can affect image quality or dose.

5.2.6.7. Corrective actions

If the device can measure the generator's kVp and dose waveform, these need to be monitored. Be aware that line voltage supply fluctuations or an electrical malfunction could influence the readings. Should the issue remain after retesting, it is advisable to reach out to the X ray system's supplier for assistance.

Should a steady discrepancy from the reference output value be detected, recalibration of the generator might be required. Conversely, if the variation is erratic, it could indicate an electrical issue within the generator. To check whether the generator is working properly, other tests need to be performed (i.e. reproducibility of the radiation output, kV accuracy, HVL and total filtration measurements).

No action is required for exposure times, but records ideally should be kept for the deviations.

5.2.7. HVL

5.2.7.1. Description and objective

The quality of the X ray beam may be described by the kVp and the HVL. This test is intended to assess the X ray beam quality to ensure accuracy and that the X ray beam is sufficient. Low X ray beam quality will lead to a higher skin entrance dose relative to a higher X ray beam quality and may be an indication that permanent filtration within the X ray tube or collimator housing has been removed or displaced.

5.2.7.2. Equipment

An appropriate radiation measurement device that directly provides an estimate of the X ray beam HVL, such as a solid state multimeter, is required. Alternatively, an ionization chamber with a relatively thin pure aluminium sheets may be used. Regardless of the measurement device used, it needs to be calibrated for the X ray beam spectra being measured.

5.2.7.3. Procedure

(a) Centre the measuring instrument within the X ray field, free-in-air if possible.
(b) Collimate the X ray field as small as reasonably achievable while ensuring that the measuring device is completely irradiated.
(c) For this test, it is preferable to use a mode of imaging allowing for manual control of the generator settings (kVp, mA, s, focal spot) and, for those system with dynamic filtration, it is important that the filtration remains constant, otherwise the test may not yield appropriate results.
(d) If a solid state multimeter with direct measurement of HVL is used:
    (i) Adequately protect the image receptor with lead attenuator.
    (ii) Set the kVp and mAs, if manually selectable, and make three exposures in the radiographic mode at approximately 60 kVp, 80 kVp, 100 kVp and a low mAs (e.g. 2 mAs).
    (iii) If the system does not allow for manual selection of the kVp and mAs, then HVL may be assessed in a clinical protocol using attenuation to drive the system close to the target kVp values indicated in step (b) above.

(e) If an ionization chamber with aluminium filters is used:
   (i) Set the kVp and mAs, if manually selectable, make exposures in the radiographic mode at approximately 60 kVp, 80 kVp, 100 kVp and a low mAs (e.g. 2 mAs).
   (ii) Record the measured air kerma.
   (iii) Add some amount of aluminium filtration, such as 2 mm, between the X ray tube and the measurement device.
   (iv) Take another exposure at the same radiographic technique (kVp, mAs, Cu).
   (v) Record the measured air kerma, if the value is more than half of that measured in step (b) above, then add slightly more filtration and repeat until the measured air kerma is less than half. If the value is less than half of that measured in step (b), then remove a small amount of filtration and repeat until a measurement is obtained slightly above half of the value in step (b). The goal is to have an air kerma measurement just slightly greater than half of the air kerma measured in step (b) without filtration and a second measurement just slightly less than this value.

5.2.7.4. *Analysis and interpretation*

The HVL provided by the solid state measurement device is straightforward, however, caution ought to be used when relying on these values. If the indicated HVL is near the limit of a tolerance, the traditional method using an ionization chamber and aluminium filters is preferred. Ionization chambers ideally should have a relatively flat response to differences in beam qualities relative to solid state devices.

When using the ionization chamber measurement, the values needed are the initial air kerma value without added filtration (I_0), the aluminium thickness (x_1) and air kerma (I_1) for a measurement of air kerma that is above 50% of I_0 as well as the aluminium thickness (x_2) and air kerma (I_2) for a measurement of air kerma that is below 50% of I_0. The traditional log-linear interpolation may be used, but it has been shown that a Lambert W interpolation may yield more accurate values [48]. The latter of these two may be determined by the following formulas:

$$\mathrm{HVL} = \frac{\ln(2)}{\mu_0 + \frac{\lambda}{2}} \tag{9}$$

where:

$$\mu_0 = \frac{x_1\left(\frac{I_1}{I_0}\right)\ln\left(\frac{I_2}{I_0}\right) - x_2\left(\frac{I_2}{I_0}\right)\ln\left(\frac{I_1}{I_0}\right)}{x_1 x_2\left(\left(\frac{I_2}{I_0}\right) - \left(\frac{I_1}{I_0}\right)\right)} \tag{10}$$

and

$$\lambda = \frac{x_2\ln\left(\frac{I_2}{I_0}\right) - x_1\ln\left(\frac{I_2}{I_0}\right)}{x_1 x_2\left(\left(\frac{I_2}{I_0}\right) - \left(\frac{I_2}{I_0}\right)\right)} \tag{11}$$

Note: For published reference and derivation see Ref. [48].

5.2.7.5. Baselines and tolerances

Requirements for minimum radiation beam qualities are often prescribed by regulatory, standards or accreditation bodies. Therefore, any such applicable sources for local or national tolerances ought to be checked first. However, even when prescribed, most tolerances do not include beam qualities incorporating spectral filtration. In these cases, other published results may be consulted for a reference [49].

Deviations from baseline data acquired during acceptance testing on the same unit using the same testing methods and devices ought to be within ±10%. Deviations greater than ±10% are to be investigated.

5.2.7.6. Frequency

During acceptance testing, annually and after major repair or replacement of the X ray tube, generator or collimator.

5.2.7.7. Corrective actions

If the measured HVL does not initially comply with the specified national standards, repeat the test and verify the accuracy of the instrument, the set-up geometry and the suitability of the attenuators. Should these tests reaffirm the discrepancy, then proceed to assess the accuracy of the kVp (Section 5.2.4) and the radiation output (Sections 5.2.5 and 5.2.6). In cases where the HVL is excessively low, it may be necessary to introduce additional, non-removable filtration, though this could lead to accelerated wear of the X ray tube. Conversely, if the HVL exceeds expectations, inspect for any removable extra filtration in the X ray beam. Various factors may contribute to changes in the HVL.

5.2.8. Accuracy of dose display

5.2.8.1. Description and objective

$K_{a,r}$, P_{KA} and PSD are indices that can be used as surrogates for patient dose. P_{KA} is an indicator of the stochastic risk, $K_{a,r}$ and PSD are indices that relate more closely to patient skin dose. The objective of the test is to check that these values are accurate and reproducible. Testing procedures are provided for assessing the accuracy of the $K_{a,r}$ and P_{KA}. If a given system provides PSD, it is crucial to consult the manufacturer's user manual to gain a comprehensive understanding of the factors incorporated into this measurement.

The PSD is a complex quantity that can encompass a range of parameters, including the area of exposure, duration of the procedure, and specific equipment characteristics. Each manufacturer may have a unique methodology for calculating PSD, taking into account different variables and assumptions. Consequently, understanding the specific calculation method is essential for accurately interpreting the PSD values. Equipment manufacturers define this quantity differently, and therefore one single approach to assessing the accuracy is not currently feasible. What is important to note is that given the variability in how different manufacturers define and calculate PSD, the medical physicist ideally should proactively request and receive detailed explanations about the PSD calculation methodology from the manufacturer.

5.2.8.2. Equipment

Air kerma rate and air kerma can be measured using a solid state multimeter. Field size can be measured with a self-developing radiochromic film or a lead ruler.

5.2.8.3. Procedure

(a) Place the measurement device on the procedure table or on the image receptor facing the X ray tube. Note that the measurement needs to be done at the reference point or it needs to be corrected using the inverse square distance law.

(b) The size of the X ray field needs to be determined in the plane of air kerma measurement. This can be achieved by placing radiochromic film on the table or on the detector or by using radiopaque rulers to measure the field size. If the measurements are made in another plane, an inverse square correction needs to be applied.

(c) Make sure that the exposure parameters are matched to the sensitivity of the radiochromic film or detector used; some radiochromic films require a very high dose to be able to measure the actual field size. If an ionization chamber is used, backscatter ought to be considered, and both detector and film (or image receptor) need to be positioned at least 20 cm from the table on polystyrene blocks.

(d) Air kerma (or air kerma rate and time) is recorded for fluoroscopy and acquisition modes.

(e) X ray voltage ought to be in the range of 90-100 kVp. Repeat exposure four times and record the results.

5.2.8.4. Analysis and interpretation

(a) If using radiochromic film, determine the field size using a ruler (or appropriate calibrated computer measuring tool).

(b) Calculate the measured P_{KA} by multiplying the air kerma results recorded by the measurement device by the X ray beam area in the plane of measurement.

(c) Calculate the deviation between the measured and displayed values. Ensure that all measured units are properly adjusted and converted for accurate comparison between the readings from the P_{KA} meter and the calibrated values from the reference instrument.

(d) The accuracy of $K_{a,r}$ and P_{KA} can be calculated using the following equation:

$$\frac{\text{(measured value - displayed value)}}{\text{displayed value}} \times 100\% \tag{12}$$

The reproducibility of $K_{a,r}$ and P_{KA} can be calculated from the four acquired measurements using the following equation:

$$\frac{\text{(measured value - mean value)}}{\text{mean value}} \times 100\% \tag{13}$$

5.2.8.5. Baselines and tolerances

It is expected that the P_{KA} meter's uncertainty will not be higher than ±25% [50]. If the measurement here will be used for both for $K_{a,r}$ and P_{KA} estimation, then tolerances are ±35% [9]. The reason is that IEC 60580 [50] refers to the P_{KA} meter alone and does not account for the field size estimation needed to estimate $K_{a,r}$. If the system has an indicator instead of an individual P_{KA} meter, then its accuracy needs to be better than ±25% [50]. The reproducibility of the repeated measurements will have a coefficient of variation of less than 0.05 (5%).

5.2.8.6. Frequency

Annually and after major repair or maintenance that can affect image quality or dose.

5.2.8.7. Corrective actions

If the P_{KA} meter's uncertainty exceeds ±25%, a correction factor needs to be applied to clinical radiation dose indices for archiving or analysis purposes. If the deviation is outside acceptable tolerances or shows inconsistency, recalibration of the meter may be necessary. Although the X ray unit can continue to operate if the performance of other components remains satisfactory, it is crucial that the P_{KA} meter is repaired as soon as possible. If the $K_{a,r}$ uncertainty is greater than ±25%, a correction factor may be introduced. Again, if these measurements deviate beyond tolerances or are inconsistent, the meter needs to undergo a new calibration immediately.

5.2.8.8. Clinical considerations

For radiation dose analysis or comparison to DRLs, correction factors need to be applied to the radiation dose indices for deviations more or less than ±10%. Otherwise, deviations between units can compound uncertainty presenting extraneous errors.

5.2.9. Image receptor entrance surface air kerma rate (for image intensifier systems)

5.2.9.1. Description and objective

The objective of this test is to verify that the air kerma rates at the entrance of the image receptor are within the acceptable limits. It is important to note that, in the case of image intensifiers, the input phosphor is subject to gradual degradation over time. This degradation may necessitate adjustments to the input dose rate to maintain image quality. However, this gradual degradation does not occur in FPDs. In FPDs, failure typically occurs through mechanisms other than slow degradation. As a result, this specific test is not necessary for systems equipped with FPDs.

5.2.9.2. *Equipment*

- A calibrated ionization chamber or solid state detector;
- 2 mm Cu plate or 20 cm thick PMMA (as attenuator).

5.2.9.3. *Procedure*

(a) Place the Cu filter on the X ray tube collimator to simulate patient thickness. Alternatively, 20 cm thick PMMA that covers the image receptor can be used.
(b) Position the image receptor to the maximum SID, or to the SID that is commonly used in clinical setting.
(c) Note: If an antiscatter grid can be easily removed, then this test is better to be done without the grid to allow for the entrance surface air kerma rate to the image intensifier to be evaluated.
(d) Adjust the collimators to match the size of the image receptor.
(e) Position the radiation detector as close as possible to the entrance surface of the image receptor or secure it directly on the image receptor surface, with tape taking care that the chamber is placed outside of the active area of the AERC (if possible). Record the focal spot to radiation detector distance.
(f) Utilize a clinical imaging protocol, apply fluoroscopy and record the air kerma rate.
(g) Conduct the fluoroscopy for all available field sizes (ensuring the entire detector is visible on the monitor) and for different dose rate settings and imaging protocols commonly used in clinical practice.
(h) Use the inverse square law to adjust all measurements for distance in order to calculate the dose rate at the image receptor if the dosimeter is not attached directly to it.

5.2.9.4. *Analysis and interpretation*

Apply the necessary correction factors to determine the entrance surface air kerma rate at the image receptor surface, accounting for elements like attenuating layers, the antiscatter grid, the table and the AERC sensor. The transmission factors are typically provided in the user manual.

5.2.9.5. Baselines and tolerances

The entrance surface air kerma rate at the image receptor in fluoroscopy mode ranges from 0.2 to 1 μGy/s for an applied tube voltage of 70–80 kVp, after accounting for grid attenuation [44]. The deviation from the established baseline values ought to not be more than ±25% (remedial level) [42] and ±50% (suspension level).

5.2.9.6. Frequency

Annually.

5.2.9.7. Corrective actions

Check the test set-up, exposure parameters and calculations, and then repeat the test. If the issue persists, contact the service engineer to do a recalibration of the detector. Depending on the deviations, the system may be limited for use or totally suspended until the recalibration or repair is done.

5.2.9.8. Clinical considerations

Image receptor entrance surface air kerma rate needs to be tested and verified to ensure that the radiation dose is optimized for adequate image quality with practically low patient dose.

5.2.10. Leakage radiation

5.2.10.1. Description and objective

This test evaluates the amount of radiation leaking from the X ray tube or collimator assembly. Radiation leaking from the tube assembly from any side other than the directed X ray beam is defined as tube leakage radiation [44].

5.2.10.2. Equipment

- Measuring tape;
- A lead sheet of appropriate thickness (e.g. 4 mm) that is able to cover the collimator exit window;

— A pressurized ionization chamber capable of measuring low ambient radiation dose rates is necessary.

5.2.10.3. Procedure

(a) Position the X ray tube so that survey measurements may be acquired in all directions such that values can be determined at 1 m from the X ray tube focal spot.
(b) Close the collimator blades completely or collimate the aperture to the smallest size available. Position the lead sheet over the aperture or exit window of the X ray tube to block any radiation directed to the image detector while allowing the system to still produce X rays.
(c) Use fluoroscopy mode with automatic control and check that the loading parameters correspond to the highest tube loading scenarios that is possible for the machine.
(d) Position the survey meter at 1 m from the X ray tube focal spot at the level where the X ray tube joins the collimator assembly (see Fig. 34).
(e) Perform measurements at multiple positions (reposition using step 4) and record the maximum air kerma rate measured for each position. The number of measurement positions can be determined on the basis of the directionality of the survey meter used. At a minimum, it should include the anode and cathode sides of the X ray tube and take into account likely positions where personnel may stand in relation to the tube assembly [44].

5.2.10.4. Analysis and interpretation

The maximum reading in any direction ideally should be used for the purposes of evaluation. Conversion of the measurement quantity may be necessary.

5.2.10.5. Baselines and tolerances

The maximum air kerma rate needs to be less than 1 mGy per hour at 1 m from the focal spot [32].

5.2.10.6. Frequency

This test is necessary at the time of acceptance, following significant repairs or when the X ray tube or collimator assembly has been replaced, as well as on an annual basis.

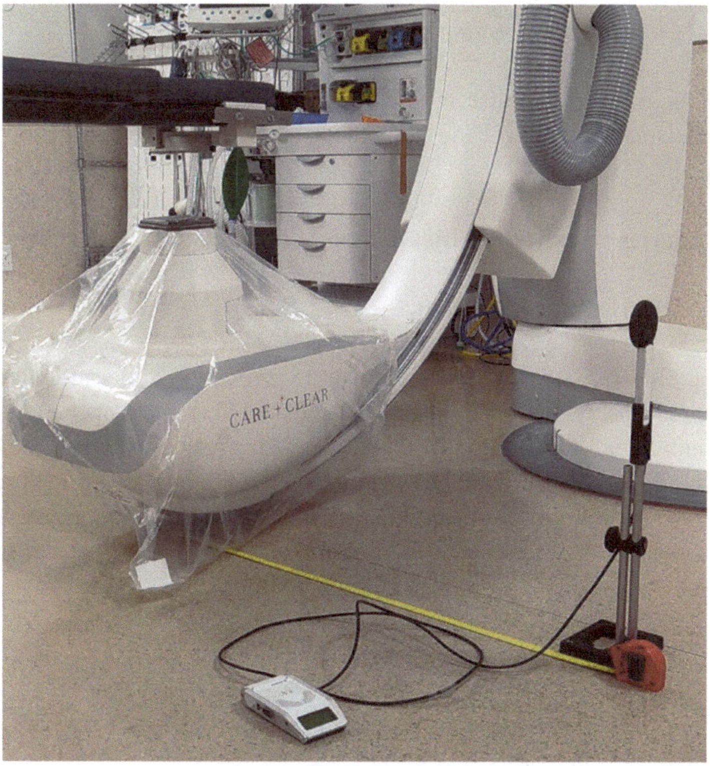

FIG. 34. Leakage radiation measurement set-up.

5.2.10.7. Corrective actions

If radiation leakage exceeds the >1 mGy/h tolerance level, it means that the radiation protection design features in the X ray tube or housing are not functioning correctly. This may lead to unnecessary exposure of patients or occupational exposure to radiation and, therefore, it is not good practice to use a device when this parameter exceeds the defined tolerance level.

Ideally measurements are repeated, and correlation of manufacturer specified tube loading parameters are reviewed prior to suspending service. Post acceptance, major service interventions in the X ray tube, housing or collimator assembly usually cause changes to this parameter.

5.2.10.8. Clinical considerations

Radiation leakage affects the radiation protection of the facility and personnel. Leakage above the tolerance levels have no effect on clinical image quality.

5.2.11. Barrier thickness verification and scattered radiation survey

5.2.11.1. Description and objective

The objective of the barrier thickness verification and scattered radiation survey is to ensure that adequate structural shielding is properly installed to protect individuals in or around the FGIP suite. Additionally, this evaluation ideally should include permanently installed shields to be used during procedures (e.g. ceiling mounted pull-down shields). Note: local or national requirements may mandate differences to the protocol described below; be sure to follow all such requirements in place of, or in addition to, the process described below.

5.2.11.2. Equipment

- Pressurized ionization chamber capable of measuring low ambient radiation dose rates is necessary;
- PMMA or other water approximating phantom of total thickness ~25 cm and sufficient surface area to cover the X ray field (~25 cm × 25 cm);
- Lead plate (~25 cm × 25 cm × 2 mm) or other highly attenuating material;
- Scintillation probe is also suggested;
- Radioactive source such as Tc-99m (optional but advised if one intends to verify actual Pb thickness installed).

5.2.11.3. Procedure

(a) Place the phantom on the procedure table and centre it within the X ray field.
(b) Orient the fluoroscopy equipment to 0 degrees of rotation and angulation.
(c) Use the largest SID possible.
(d) Use a table height expected during clinical procedures.
(e) Choose a high pulse rate (e.g. 30 pps).
(f) Choose the normal or high fluoroscopy dose curve.
(g) Place lead plate on top of the PMMA stack.
(h) Collimate the image to the edges of the lead.

If the fluoroscopy equipment terminates X ray production with the lead blocking the image receptor, pull the lead back slightly so there is an area of transmission, but be sure to cover that area with several mm of Cu.

(i) The system is now set to perform measurements:
   (i) For all permanently installed shields in the room:
      — During fluoroscopy, make ambient air kerma rate measurements with the pressurized ionization chamber at a fixed distance from the phantom, with and without the shield between scattering medium and the measurement device. The ratio of the air kerma rate measurement with the shield in place to that without will give the transmission factor.
   (ii) For structural shielding verification:
      — Using the same set-up described in (a)–(h), perform fluoroscopy while doing the following:
         - First, using a scintillation detector, survey every barrier (wall, window, door, etc.), looking for large spikes in the readings that may indicate a penetration, incomplete overlap of the shielding material, or missing shielding material. These measurement values are only relative, as a relatively fast and large response will be observed if an area of concern is encountered. Scintillation detectors are much better suited for these measurements than are pressurized ionization chambers because of the slow temporal response of the ionization chambers.
         - Once the integrity and consistency of the barrier have been confirmed, change to the ionization chamber for ambient air kerma measurements. Air kerma rate measurements need to be made behind each barrier to allow estimation of occupational radiation doses to individuals shielded by that barrier. There are several approaches to these methods; be sure to understand the appropriate methodology, applicable criteria and any local or national requirements [24, 51].

Optional: thickness verification.

The following method can be used if it is desired to assess the lead equivalence of a given barrier to either verify the shielding design requirements or to assess the value of an unknown barrier:

(a) Be sure to check with the local radiation safety officer and local regulations to ensure compliance. Take all necessary radiation safety precautions, if not familiar with working with or around radioactive material, do not attempt this procedure.

(b) Acquire a radioactive source (commonly Tc-99m) of approximately 5.6 GBq (150 mCi). The source needs to be sufficient to penetrate the barrier and provide a reading to determine the thickness. The source ought to be in an appropriate shield.

(c) Set-up the source in the room to be evaluated; it is best to have the shielded source in a vial, as opposed to a syringe. Additionally, absorbent material ought to be placed in the vial prior to introducing the radioactive material. This will limit the damage if the source was to drop or break.

(d) Place an ionization chamber approximately 2 m from the shielded source.

(e) Remove the source from its shielding and acquire an ambient air kerma rate measurement.

(f) Replace source in its shielding.

(g) Place the source in the imaging suite two m from a barrier. It is typically best to start with a door, since the distance can easily be determined.

(h) Measure or determine the distance from the source to the location where the measurement will be made, and record this value.

(i) Remove the source from its shield and acquire an air kerma rate measurement behind the barrier.

(j) Take the ratio of the air kerma measurement behind the barrier to the air kerma measurement initially made at 2 m. Be sure to distance correct (inverse square) one or the other air kerma measurements so that the values are corrected to the same distance.

(k) The value obtained by taking the ratio is the barrier transmission. Compare the transmission to published values to determine the lead equivalence of the barrier [24, 51].

(l) Repeat steps above for all the barriers.

5.2.11.4. Analysis and interpretation

For the evaluation of the in-room shields, the transmission factor needs to be less than 0.25 (25%). Typical transmission values for a 0.5 mm Pb equivalent shield ought to be approximately 0.05 (5%).

For the structural shielding survey, first obtain the regulatory and shielding design goal values for the areas being evaluated. Then determine whether the measured values meet the regulatory and design goal values for the occupational rates and annual estimations.

For the optional thickness verification, compare the calculated barrier thickness to the design requirement if the measurement was to confirm the Pb equivalence of the design.

5.2.11.5. *Baselines and tolerances*

All measurements need to be below any applicable regulatory limits and need at minimum to meet the shielding design goals.

5.2.11.6. *Frequency*

These measurements need to be performed only at equipment acceptance testing or be repeated only if any change to the barriers occur (i.e. damage, construction).

5.2.11.7. *Corrective actions*

If any barrier fails to meet the regulatory requirements, the issue ought to be mitigated. If the barriers meet the regulatory requirements but do not align with the shielding design goals, then the reason for not meeting the goal needs to be investigated.

5.2.12. Monitor display check

QC tests for monitor display include test pattern evaluation, minimum and maximum luminance and ratio, luminance response, and luminance uniformity.

5.2.12.1. *Test pattern evaluation*

This test is a qualitative evaluation of the performance of the display system.

5.2.12.1.1. Description and objective

The consistent display of a known pattern provides reassurance that the display system is functioning as expected with no obvious faults exceeding the action levels.

5.2.12.1.2. Equipment

This test requires the display of a suitable test pattern. The following test patterns are suggested for this test: AAPM TG18 QC [47], IEC modified TG18-OIQ [45], AAPM TG270 sQC and AAPM TG270 pQC [25].

The older versions of the Society of Motion Picture and Television Engineers medical diagnostic imaging test pattern is not endorsed by this report for the purposes of testing LED monitors, as it was designed for older CRT monitors not covered in this report.

5.2.12.1.3. Procedure

(a) Load the suitable test pattern on all applicable display monitors used clinically during routine operation of the FGIP equipment. This needs to include the monitors next to the patient table as well as the control room monitors. FGIP systems with large displays that make use of customizable layouts or image views can use the supplier pixel matrix for testing purposes.
(b) Evaluate the qualitative features of the chosen test pattern. The evaluation procedure will vary depending on the type of test pattern being used, as each has different features. The CQMPs are advised to consult the original report describing the respective test patterns for detailed instructions on each evaluation.
  (i) For TG18-QC [47] and TG18-OIQ test patterns [45]:
    — Perform a qualitative evaluation of the low contrast sections to evaluate the luminance response function. Pay attention to the following sections:
      - 0%/5% and 95%/100% contrast objects;
      - 'QUALITY CONTROL' letters visible at different levels of background contrast;
      - 16 luminance patches and their smaller contrast corner patches;
      - Luminance ramps along the sides of the patterns.
    — Inspect the displayed pattern for artefacts or abnormalities.
  (ii) For TG270-sQC test pattern [25]:
    — Evaluate the luminance response and pay attention to:
      - Three rows of greyscale patches (total 18 patches) with incrementing luminance.
      - Evaluate the bar pattens in the upper left and lower right corner of each greyscale patch. These bar patterns are +/-5 grey levels different from the background.
    — The greyscale gradient at the bottom of the pattern.
    — Inspect the displayed pattern for artefacts or abnormalities.

5.2.12.1.4. Analysis and interpretation

The analysis and interpretation of this test is qualitative in nature and therefore relies on and assumes normal functioning of the human visual system of the CQMP or the expert operator performing the evaluation. Nonetheless, this qualitative test is a valuable tool to evaluate the correct functioning of the display monitors because it is relatively quick to perform. Quantitative tests, performed using a photometer, are more accurate than the human visual system and ideally should be considered for more advanced testing or when the qualitative tests highlight possible problems.

5.2.12.1.5. Baselines and tolerances

The baselines and tolerances are discussed for each pattern below:

(a) For TG18-QC and TG18-OIQ test patterns:
   (i) 5% and 95% inserts ought to be visible in normal light settings;
   (ii) 'QUALITY CONTROL' letters need to be visible at the different levels of background luminance;
   (iii) All 16 luminance patches need to be visible;
   (iv) No artefacts hindering clinical use ought to be present on the display.
(b) For TG270-sQC test pattern:
   (i) All 18 greyscale patches need to be distinguishable;
   (ii) All the +/-5 grey level bar patterns inside of the 18 greyscale patches need to be visible;
   (iii) No artefacts hindering clinical use ought to be present on the display.

5.2.12.1.6. Frequency

Annually or when the MRT monitor check (described in Section 5.1.3) reaches tolerance levels needing a CQMP investigation.

5.2.12.1.7. Corrective actions

Repair or replacement of the affected display ideally should be performed by the equipment manufacturer. Facilities with enough multiple displays may consider restricting the clinical use of the affected display so that the interventional system may still be used while the repair or replacement is underway.

5.2.12.1.8. Clinical considerations

The incorrect functioning of the image display monitors of an FGIP system, outside of the tolerance levels, can impact the user's ability to visualize and interpret the images correctly during procedures. This may lead to missed diagnoses or even mistakes during interventions performed that can negatively affect the clinical outcome of patients.

5.2.12.2. Minimum and maximum luminance and ratio

This test evaluates the display luminance intensity by measuring the minimum luminance, maximum luminance and the ratio thereof.

5.2.12.2.1. Description and objective

The QC of display luminance intensity needs to be considered together with that of the luminance response function test described in Section 5.2.12.3 because the observer's experience of brightness and contrast are dependent on them both. The following descriptions are used to evaluate display luminance intensity:

Ambient luminance (L_{amb}) describes the amount of ambient light that is reflecting from the surface of the display and is measured in cd/m^2. L_{amb} may be determined by direct measurement using a photometer, which will be described in this report, but may also be determined through calculation or estimation [25, 52].

Minimum luminance (L_{min}) describes the intensity of the lowest grey value on the display and is measured in cd/m^2. L_{min} on displays ought to be set so that the ambient ratio (AR) is defined as L_{amb}/L_{min} is < ¼ [25]. The total minimum luminance (L'_{min}) describes the minimum luminance as viewed by the operator as it considers the added ambient luminance component. L'_{min} is calculated by adding $L_{min} + L_{amb}$.

Maximum luminance (L_{max}) describes the intensity of the highest grey value on the display also measured in cd/m^2. The total maximum luminance (L'_{max}) considers the added ambient luminance and is calculated as $L_{max} = L_{max} + L_{amb}$.

The luminance ratio (LR) is defined as the ratio of L'_{max}/L'_{min} and is usually set to target approximately a ratio of 350 (±100).

5.2.12.2.2. Equipment

A calibrated photometer (a telescopic photometer is preferable) able to measure luminance (cd/m^2) and a suitable test pattern displaying minimum and maximum greyscale levels [47, 52].

5.2.12.2.3. Procedure

This test ideally should be performed on all displays used clinically in the FGIP operating room and control areas. L_{amb} needs to be determined first and can be measured directly using a photometer. This is done by switching off the display that is being tested so that no light is produced by the display backlight and only ambient reflected light is measured. The measurement of L_{amb} needs to be done with FGIP room lighting set to typical clinical conditions. Place the photometer in a viewing direction that a clinical observer would typically use, then allow the reading to stabilize and measure L_{amb} in cd/m^2 [25].

Following the determination of L_{amb}, the minimum and maximum luminance ought to be measured. L_{min} is measured by displaying the appropriate test pattern with the lowest grey level (0) and then measuring the luminance using the photometer. L_{max} is measured similarly to L_{min}, except that the pattern used needs to display the maximum available grey level. L_{amb}, L_{min} and L_{max} ought to be recorded for every monitor being evaluated.

5.2.12.2.4. Analysis and interpretation

Using the formulas below, calculate AR and LR for each display being evaluated:

$$\mathrm{AR} = L_{amb} / L_{min} \text{ and } \mathrm{LR} = L'_{max} / L'_{min} \tag{14}$$

5.2.12.2.5. Baselines and tolerances

(a) AR < ¼;
(b) LR = 350 (250–450);
(c) $L'_{min} \geq 0.8$cd/m^2;
(d) $L'_{max} = 250$ cd/m^2 (≥ 200 cd/m^2).

5.2.12.2.6. Frequency

Annually.

5.2.12.2.7. Corrective actions

The following can be considered as corrective measures:

(a) Reduce room illuminance by dimming the lights used clinically, as this will improve AR.
(b) The affected display can be positioned away from reflecting lights, possibly increasing L_{amb}.
(c) Displays may be repaired or replaced with higher L_{max} models.

It is important to note that the ambient lighting and resulting illuminance in FGIP operating rooms is normally brighter than that of general radiology reporting rooms, which in some cases will make the proposed AR and LR difficult to achieve. Especially if the display model has a L_{max} in the lower range (~250), maintaining an AR < ¼ in a bright environment would require that L_{min} be increased considerably. This adjustment is not preferred, as it could result in a loss of contrast at low luminance levels if the lighting is dimmed during procedures. Consultation with the specialist doctor on clinical acceptability is advised [52].

5.2.12.2.8. Clinical considerations

Incorrect ARs and LRs may cause a clinically relevant loss of contrast, especially low contrast detail in lower grey levels, if the display set-up is not correctly set-up initially and maintained.

5.2.12.3. Luminance response

The DICOM standard Grey scale display function (GSDF) is a widely accepted standard designed to ensure that the luminance response of medical displays is optimized for the human visual system. Despite being a non-linear luminance curve, the difference in luminance required for an observer to be noticeable, or the just noticeable difference, ought to be linear from the observer's perspective when the GSDF is correctly applied. The luminance response curve of FGIP modality displays better be standardized according to the DICOM GSDF [46, 53].

5.2.12.3.1. Description and objective

This test evaluates whether the luminance response of the display is responding according to the DICOM GSDF at different grey levels between L_{min} and L_{max}. The quantitative version of this test is intended to be performed on displays used in the procedure room for diagnostic interpretation or interventional procedure guidance. The qualitative version of this test is intended for the remaining monitors that may be used to displays in the imaging suite.

5.2.12.3.2. Equipment

Calibrated photometer able to measure luminance (cd/m^2) and a suitable test pattern or set of patterns to display varying luminance levels. The TG18-LN, TG270-ULN and TG270-sQC test patters are appropriate for quantitative assessment. Qualitative assessment may be done using the TG18-QC, TG18-OIQ, TG270-sQC or TG270-pQC.

5.2.12.3.3. Procedure

The quantitative evaluation of luminance response function can be applied using at least the 18-point or 52-point techniques described in the AAPM TG18 [15] and AAPM TG270 [14] reports. Depending on the patterns available on the FGIP equipment, the reader can refer to the AAPM TG270 report (Section 2.3.2) for an in-depth description of the DICOM GSDF testing procedure and required calculations [25, 47].

The qualitative evaluation of luminance response has limitations owing to the inherent inaccuracies of the human visual system when compared to photometer measurements and it is unlikely that an observer would detect small deviations for the DICOM GSDF using these qualitative techniques.

5.2.12.3.4. Analysis and interpretation

The analysis of the 18-point or 52-point techniques is best shown graphically by plotting the measured luminance (cd/m^2) and grey levels in relation to the DICOM GSDF standard function. The luminance response needs to be evaluated over the entire luminance range available on the display between L_{min} and L_{max} and to conform to within the set tolerance levels.

5.2.12.3.5. Baselines and tolerances

Deviation ≤ ±20% from the DICOM GSDF.

5.2.12.3.6. Frequency

Annually.

5.2.12.3.7. Corrective actions

Recalibration of the display according the DICOM GSDF standard. If recalibration is not successful, replacement of the effected display ought to be considered.

5.2.12.3.8. Clinical considerations

The luminance response of the display has a direct clinical impact on the information available to the user, and faulty displays can lead to missing or incorrect information observed. When the luminance response of a display is correctly calibrated according to the DICOM GSDF, it ensures that the display has adequate contrast available over the entire luminance range available between L_{min} and L_{max}. It also creates a linear just noticeable difference response, which guarantees that the luminance response function and the information being displayed are optimized for the human visual system to detect.

5.2.12.4. Luminance uniformity

This test verifies that the luminance output is uniform across the display area.

5.2.12.4.1. Description and objective

The quantitative version of this test is intended to be performed on displays used in the procedure room for diagnostic interpretation or interventional procedure guidance. The qualitative version of this test is intended for the remaining monitors that may be used as displays in the imaging suite.

Qualitative evaluation of displays is preferred on FGIP systems, as their displays vary in size, and some systems have extremely large displays that would normally be configured to display various video inputs and have configurable layouts. Clinical use and relationship to the position of the display area being evaluated need to be considered when assessing large displays, as they may not be able to maintain uniformity across the entire panel due to their size [25].

Quantitative evaluation of luminance uniformity may be considered for acceptance testing or problem-solving scenarios. This is performed using a photometer and calculating the luminance uniformity deviation from the median. If required, the reader can refer to the AAPM TG270 report [14], where the 9-point luminance uniformity deviation from the median quantitative test is described in detail.

5.2.12.4.2. Equipment

A series of uniform test patterns that can display varying levels of luminance is required. The TG18-UN10, TG18-UN80, TG270-ULN8-200, TG270-ULN8-100 and TG270-ULN8-020 are suitable test patterns for this test.

5.2.12.4.3. Procedure

Luminance uniformity needs to be evaluated at low, medium and high luminance levels by displaying the mentioned test patterns on each of the displays used clinically. Large displays ought to be evaluated in the commonly used clinical layout, as it is not feasible to evaluate all possible configurations.

Evaluate the pattern from the typical clinical observer's perspective and search for any large or small non-uniformities that appear different from the rest of the display. Ageing LCD displays commonly suffer from faults in the LCD alignment layer, called 'mura', that can cause light or dark blotches on the display. The display also needs to be evaluated for any bad pixels, which can present as dark or bright individual pixels.

5.2.12.4.4. Analysis and interpretation

The impact of any non-uniformities that have been found should ideally be analysed in the context of the location, frequency of occurrence and behaviour over the entire LR. FGIP displays fall into the category 'modality displays', which may develop subtle non-uniformities (or 'mura') or bad pixels as the display ages. Although not ideal, subtle non-uniformities may still be acceptable, as the non-uniformity will not likely have any clinical impact. The CQMP needs to consult with the clinical users when interpreting whether an identified non-uniformity may have a clinical effect or not.

5.2.12.4.5. Baselines and tolerances

Luminance variations evaluated qualitatively are ideally unnoticeable and not obvious when evaluating any of the low, medium or high luminance test patterns. Any luminance variations that exist ideally ought to be clinically acceptable.

5.2.12.4.6. Frequency

Annually.

5.2.12.4.7. Corrective actions

FGIP displays that have non-uniformities affecting clinical use need to be repaired or replaced.

5.2.12.4.8. Clinical considerations

Non-uniformities that fall within the clinically used part of the FGIP display may affect the user's ability to interpret diagnostic information correctly and may cause the user difficulty when navigating instruments during interventions.

5.2.13. Artefact evaluation

5.2.13.1. Description and objective

Image artefact assessment is a process of evaluating imaging studies to identify any artefacts that may have been introduced during the imaging process or any flaws in the image caused by certain conditions of the environment or the scanning device. This assessment is used to determine the accuracy and reliability of the image and can help identify factors that may contribute to a misdiagnosis or incorrect conclusions from an image. The assessment can also help ensure that radiological medical practitioners and other observers accurately read and interpret the image to make an accurate diagnosis.

5.2.13.2. Equipment

Grid patterns with different spacings.

5.2.13.3. Procedure

Expose the patterns using fluoroscopy mode.

5.2.13.4. Analysis and interpretation

Verify that the displayed images conform to the test patterns.

5.2.13.5. Baselines and tolerances

No clinically relevant artefacts are allowed. For image intensifier based systems, pincushion or S-distortion is expected for large FOVs (Fig. 35).

5.2.13.6. Frequency

Annually and after major repair or maintenance that can affect image quality.

5.2.13.7. Corrective actions

Action level: If clinically relevant artefacts are detected, first seek and potentially resolve the source of the artefact. Otherwise, call for service. Examples of artefacts are shown in Figs 36–38.

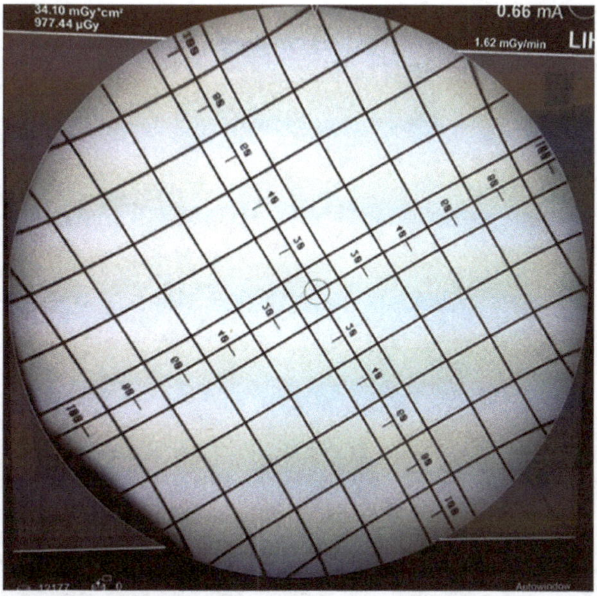

FIG. 35. S-distortion artefact from the image intensifier.

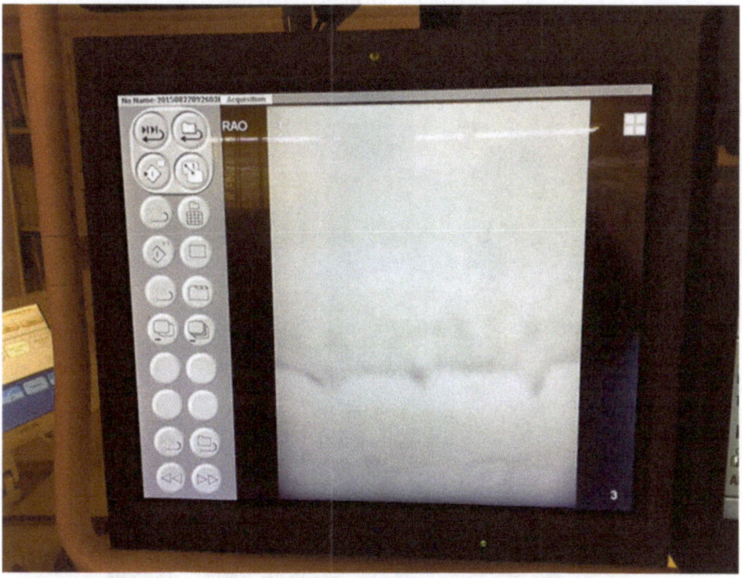

FIG. 36. Artefact in a blood infiltrating collimator assembly.

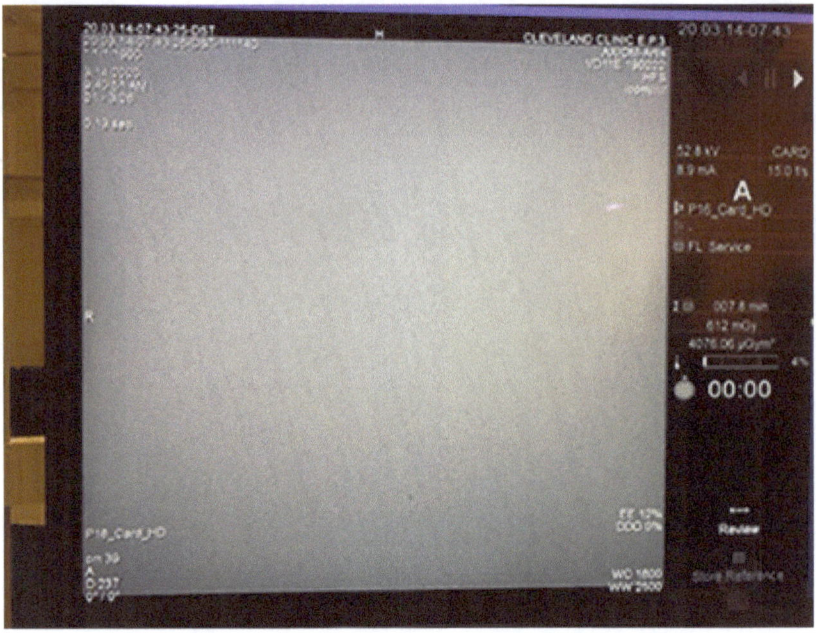

FIG. 37. Grid artefact.

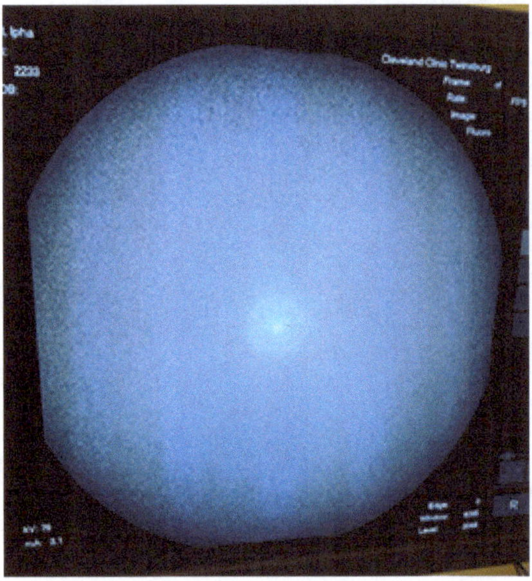

FIG. 38. Gassing artefact from an image intensifier.

5.2.14. Technical versus clinical image quality assessment

To emphasize the importance of understanding both quantitative and qualitative aspects of image quality, this section focuses on technical image quality assessment and its distinction from clinical image quality observed during patient procedure. Technical image quality assessment is necessary to quantitatively assess measurable parameters. However, this is not always directly relatable to clinical image quality observed in patients. Clinical image quality is more complex, and advanced image processing may not allow static phantoms without clinically relevant anatomy to adequately describe what is observed clinically. Technical image quality assessment ought to include low contrast visibility (threshold detectability) and high contrast visibility (limiting spatial resolution).

Before image quality assessments are performed, the display monitor evaluation tests need to be completed to ensure proper functioning of the display(s). This preliminary step guarantees that all assessments are conducted under optimal conditions, providing accurate and reliable results that are crucial for effective patient outcome.

5.2.15. Low contrast visibility

5.2.15.1. Description and objective

Low contrast visibility is a crucial indicator of image quality in fluoroscopy and FGIPs, denoting the system's capacity to visualize small objects with minimal contrast [54]. It is primarily dependent on contrast, sharpness and background noise [55]. There are several methods to assess low contrast visibility, which can be generally divided into quantitative and qualitative assessments. For qualitative assessment, a standard test object (e.g. Leeds test object N3 or similar) is commonly used to have a subjective evaluation of the visibility of various low contrast details by human observers. For quantitative assessment, a commonly used parameter to estimate low contrast visibility is known as accumulation rate of SNR^2 (SNR^2_{rate}). SNR^2_{rate} increases linearly with the area of the contrast detail (A) and incident air kerma rate (K), which is in agreement with the so-called Rose-model, $SNR^2_{rate} \propto M^2 C^2 A.K$, where C is the contrast and M is the magnification [55]. Details regarding this approach can be found in relevant literature [56, 57]. Considering time and practicability, qualitative assessment through visual inspection is usually preferred in routine QC. However, the test requires a baseline to be set (if possible, during commissioning) [44]. Any deviation from the baseline may indicate a change in imaging performance.

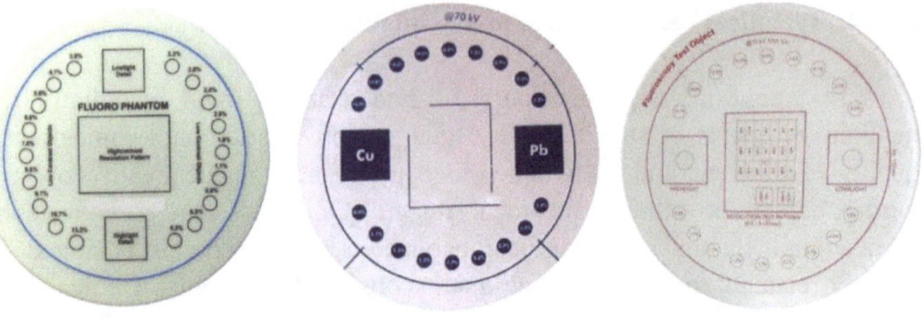

FIG. 39. Examples of commercially available low contrast test object.

5.2.15.2. Equipment

Most phantoms used for image quality assessment were designed several decades ago, before the advent of FPD image receptors. Caution ought to be used in using the associated tolerances, as the results may not be equivalent.

The following equipment is required:

— Low contrast test object (examples shown in Fig. 39) and the manufacturer's user manual;
— 1 mm copper plate or other attenuators.

5.2.15.3. Procedure

(a) Place the low contrast test object on the table as close to the image receptor as possible (see Fig. 40).
(b) Adjust the distance of the SID to 100 cm, if possible (in mobile systems this is not always feasible).
(c) Place the 1 mm copper plate or other attenuator on the X ray tube assembly.
(d) Make a fluoroscopy exposure with AERC or follow the test object's manufacturer specification (e.g. 70 kVp with AERC).

5.2.15.4. Analysis and interpretation

Conduct a visual assessment of the LIH on the display, noting the count of low contrast objects that can be resolved in the image. Document this number and compare it with the established baseline. This straightforward test, rooted in subjective visual inspection, aims to identify any trends indicating a decline in the performance of low contrast imaging [44].

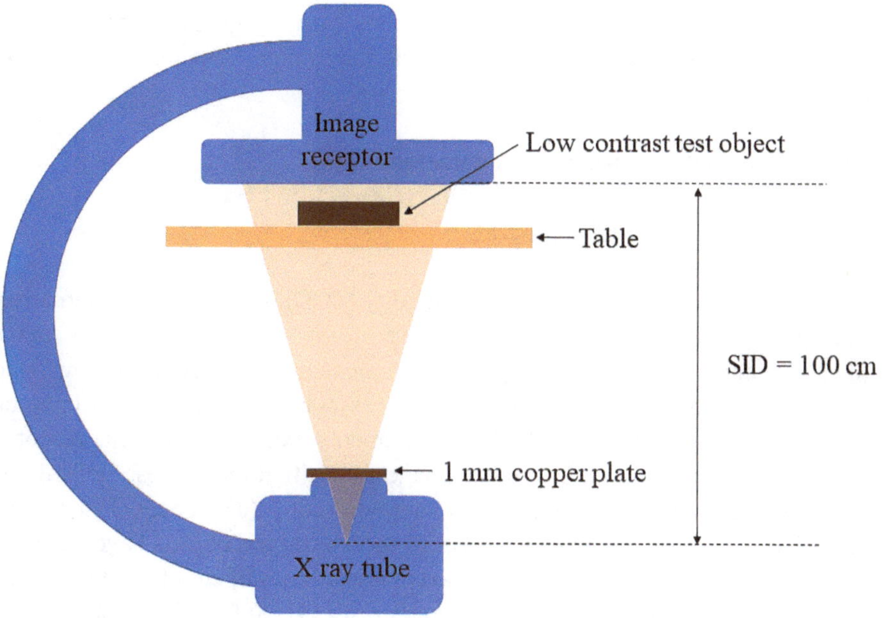

FIG. 40. Test set-up for low contrast visibility.

5.2.15.5. Baselines and tolerances

According to the test object used, the tolerance from the baseline value ought to be decided and proposed by the CQMP in cooperation with the FGIP radiological medical practitioner and the MRT.

5.2.15.6. Frequency

Annually.

5.2.15.7. Corrective actions

If the tolerance limit is exceeded, first check the positioning of the test object and attenuator, make sure that the test object is within the X ray field and the attenuator covers the entire tube output port and that AERC mode is turned on, then repeat the test. Variations in low contrast visibility could stem from alterations in the imaging protocol, test set-up or changes in radiation output and HVL. Initially, verify these elements for consistent measurements. The degradation in low contrast visibility might also be attributed to the ageing of image receptors, such as image intensifiers. Engage with users to confirm that such changes do not compromise clinical imaging quality. Depending on the degree of image quality impact, actions may range from restricting use to suspending service.

5.2.16. Threshold contrast detail detectability

5.2.16.1. Description and objective

The threshold contrast detail detectability (TCDD) test offers an alternative, more practical approach to assessing the performance of conventional image intensifier and FPD systems [58]. When implemented in a controlled manner, the TCDD test provides a simple and quick assessment of the overall imaging capabilities of the system. TCDD measurements extend beyond fluoroscopic imaging, encompassing various other image acquisition modalities. Several test objects designed for assessing TCDD in fluoroscopy (e.g. Leeds Test Object TO10, TO12, TO20) are available. Most of the TCDD test objects have a similar design and scoring methodology. They comprise a series of circular detail groups of different diameters and radiographic contrast [59]. A TCDD test object may feature either positive or negative details. Positive details consist of attenuating material discs placed on a uniform base plate, typically made of PMMA, whereas negative details are represented by holes drilled into a uniform base plate, which is often also PMMA.

The images of the test object acquired under a standard exposure setting will be evaluated visually at a fixed viewing distance. The lowest contrast detail that can be visible for each group of the same diameter will be recorded. Details can be scored as half visible if the detail is partially distinguishable [59]. TCDD data can be presented graphically as a contrast resolution diagram or more commonly as the threshold detection index (H_{T}) calculated using the following equation [59]:

$$\mathrm{H}_T(A)=\frac{1}{\mathrm{C}_T\times\sqrt{A}} \tag{15}$$

where C_{T} is the threshold contrast (%) and A is the area of the detail (cm^2).

5.2.16.2. Equipment

The following equipment is required:

- — A calibrated dosimeter;
- — Threshold contrast test object (examples shown in Fig. 41) and its user manual;

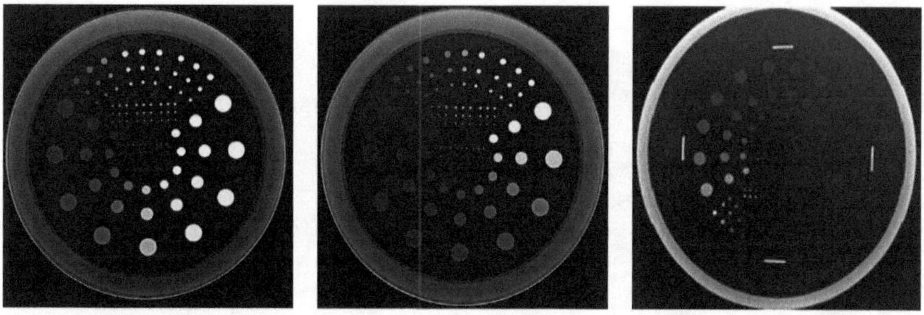

(a) (b) (c)

FIG. 41. Examples of test objects images for (a) fluoroscopy, (b) digital spot imaging and (c) DSA image quality assessment.

— 1 mm copper plate or other attenuators.

5.2.16.3. Procedure

(a) Place the TCDD test object on the table as close to the image receptor as possible (see Fig. 42).
(b) Adjust the distance of the SID to 100 cm, if possible (in mobile systems this is not always feasible).
(c) Place the 1 mm copper plate or other attenuator on the X ray tube assembly.
(d) Place the detector of the dosimeter on the image receptor surface to measure the input air kerma rate.
(e) Select the normal field size.
(f) Make a fluoroscopy exposure with AERC.
(g) Record the measurement of $\dot{K}$.
(h) Repeat steps (e) to (g) with different field sizes. If the smaller field size does not cover the entire test object, move the table or test object around the FOV to view all visible details.

5.2.16.4. Analysis and interpretation

Review the last image retained on the display, noting the least visible contrast detail for each group with identical diameters. Details need to be marked as half visible if they are only partially distinguishable. Determine the threshold contrast (%) by referring to the manufacturer's user manual. Calculate H_T using the following equation or plot a graph of H_T versus $\sqrt{A}$ for the particular field size. Compare the H_T with the baseline data.

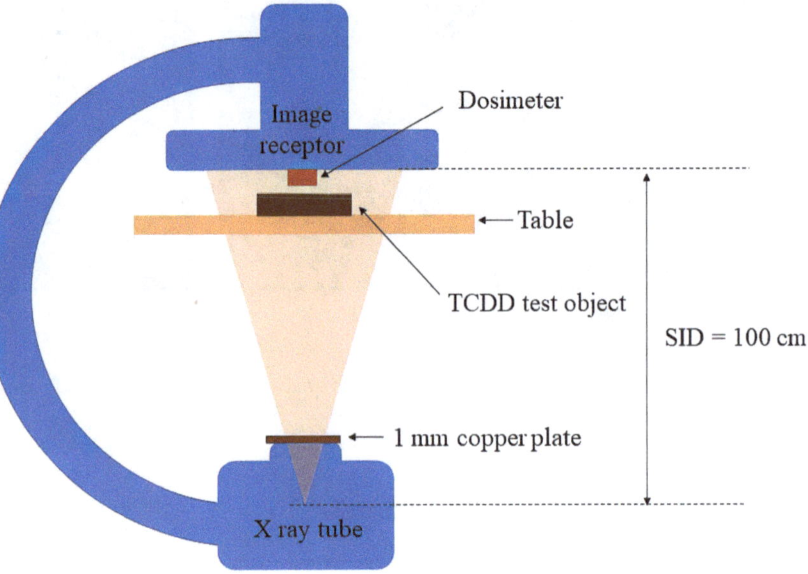

FIG. 42. Test set-up for TCDD evaluation.

Additional analysis:

Calculate image quality factor using the following equation proposed by Gallacher et al. [60]:

$$\mathrm{IQF} = \frac{1}{n} \times \sum_{i=1}^{n} \frac{H_{\mathrm{T}}(A_i)}{H_{\mathrm{Tref}}(A_i)} \sqrt{\frac{\dot{K}_{\mathrm{ref}}}{\dot{K}_{\mathrm{m}}}} \tag{16}$$

where

n is the number of detail diameter groups;
$H_{\mathrm{T}}(A)$ is the threshold detection index calculated from the image;
$H_{\mathrm{Tref}}(A)$ is the threshold detection index calculated from a reference image of a system known to be in good adjustment;
$\dot{K}_{\mathrm{ref}}$ is the input air kerma rate for the reference image (in nGy/s);
and $\dot{K}_{\mathrm{m}}$ is the measured input air kerma rate of the system (in nGy/s).

5.2.16.5. Baselines and tolerances

H_{T} or image quality factor should not exceed 10% difference from the baseline.

5.2.16.6. Frequency

Annually.

5.2.16.7. Corrective actions

In cases where the tolerance threshold is surpassed, first verify the measurement set-up and exposure settings before redoing the test. Alterations in TCDD could result from modifications in the imaging protocol, the test configuration or variations in radiation output and HVL. Assess these aspects initially in order to confirm measurement consistency. Additionally, the declining performance of TCDD might be attributed to the ageing of image receptors, particularly image intensifiers.

5.2.17. High contrast (spatial) resolution

5.2.17.1. Description and objective

This test is also referred to as limiting spatial resolution. A line pair resolution test pattern (e.g. Huttner resolution test pattern or Leeds Test Object TOR 18FG) is commonly used to assess the best spatial resolution that can be achieved by the system under a standardized set-up and exposure condition. The spatial resolution is determined by visually inspecting which groups of line pairs are clearly visible on the image. The exposure ideally should be done at a high input exposure rate to avoid noise on the image that could affect the assessment. The spatial resolution is expressed in lp/mm. This test requires a baseline to be set, preferably during commissioning [44], and there ought not to be any deterioration of the limiting spatial resolution compared to baseline.

Among all image quality tests, this test is the most sensitive to magnification, focal spot size and pixel size (for FPD systems) [52]. Therefore, care needs to be taken during the positioning of the test object to avoid aliasing with the regular pixel matrix. Rotating the bar pattern to 45° to the image receptor edge will achieve the highest possible resolution of this test. Most importantly, as previously mentioned, the test set-up, exposure parameters and evaluation conditions should maintain the same as the baseline measurement and during each of the periodical QC tests.

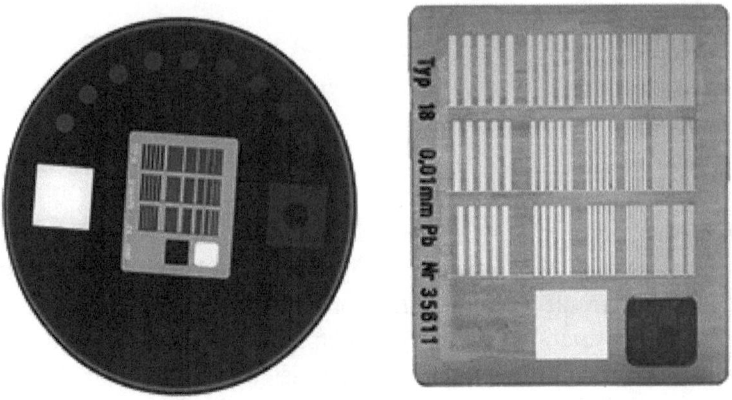

(a) (b)

FIG. 43. Examples of (a) Leeds TOR 18FG image and (b) Huttner resolution test pattern.

5.2.17.2. Equipment

The following equipment is required:

- Line pair resolution test pattern or Leeds test object TOR 18FG, or equivalent (examples shown in Fig. 43);
- 1 mm Cu or equivalent in other material (optional, depending on calibrated conditions).

5.2.17.3. Procedure

(a) Remove antiscatter grid from the system, if possible.
(b) Place the line pair test pattern or test object on the table as close to the image receptor as possible (see Fig. 44). If grid cannot be removed, rotate the resolution test pattern to 45° to the FOV.
(c) Adjust the distance of the SID at 100 cm, if possible (in mobile systems this is not always feasible).
(d) Select normal field size.
(e) Make a fluoroscopy exposure at low X ray tube voltage of 40–60 kVp. A high input exposure rate should be used to suppress noise in the image. Note: For some systems, PMMA phantom may need to be added to achieve standardized conditions.
(f) Repeat steps (d) to (e) for each available field size.

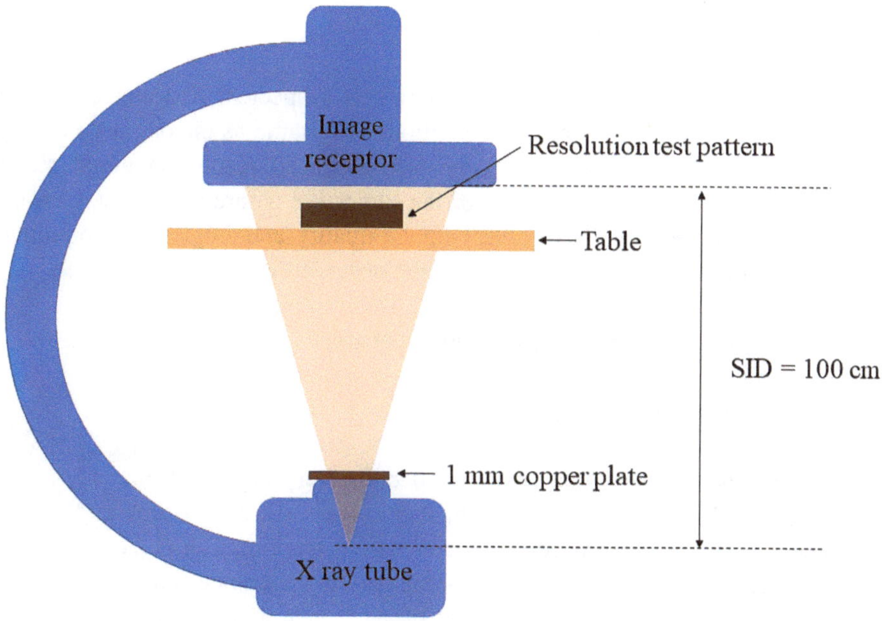

FIG. 44. Test set-up for high contrast spatial resolution measurement.

5.2.17.4. Analysis and interpretation

Inspect the most recent image on display, utilizing windowing and magnification to enhance the visibility of the test object. Identify which groups of line pairs can be seen in the image. For a group to qualify as 'visible', its line pairs must be discernible across their entire length. The final group where each bar appears as a distinct line determines the maximum spatial resolution. Look up the corresponding spatial resolution in lp/mm from the manufacturer's user manual. Compare the results with the baseline data.

5.2.17.5. Baselines and tolerances

There ought not to be any change in limiting spatial resolution from the baseline.

5.2.17.6. Frequency

Annually.

5.2.17.7. *Corrective actions*

If deterioration of the spatial resolution is suspected, first check the test set-up and make sure the exposure condition is the same as the one used during baseline measurement, then repeat the test. Reduced resolution might result from suboptimal viewing conditions or the subjective nature of the test. Consult the most recent test results to assess image quality and implement corrective measures to eliminate potential causes of the observed variation [44].

5.2.18. Patient entrance surface air kerma rate

5.2.18.1. *Description and objective*

In FGIPs, the local skin dose may become very high, and this may result in skin burns. Estimation of the patient's entrance surface air kerma rate during fluoroscopically guided procedures is therefore important. The objective of this test is to verify that the air kerma rate at the patient's entrance is under the limits for the different imaging protocols, FOV and operation modes (fluoroscopy modes, cine or acquisition modes) used under normal clinical conditions in X ray fluoroscopy systems. Apart from this purpose, the measured results are also indicative for the evaluation of expectable peak skin dose levels [3, 32, 40, 44].

5.2.18.2. *Equipment*

— Calibrated dosimetry equipment (ionization chamber or solid state detector). The dosimetry equipment ought to not affect the AERC settings.
— 20 cm thick PMMA plate. Use a 20 cm phantom comprising of rectangular blocks (for example: 5 cm, 10 cm, 15 cm, 20 cm, 25 cm and/or 30 cm PMMA) with sides equal to or exceeding 25 cm [32].

Note: A water phantom could alternatively be used in place of the PMMA [44].

5.2.18.3. *Procedure*

(a) Figure 45 shows the test set-up for this measurement in posterior-anterior (C-arm in 0° of rotation) and lateral (C-arm in 90° of rotation) projections.
(b) Check the antiscatter grid is in its right position.
(c) Position the PMMA plates on the patient's table in such a way that between the table and the first PMMA plate there is enough space to put the dosimetry equipment (other PMMA plates or spreaders can be used, as well).

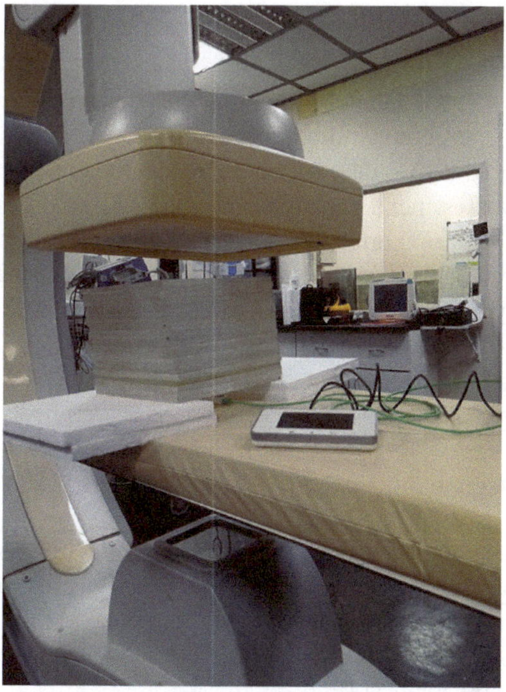

FIG. 45. Experimental arrangement for the estimation of the patient's entrance surface air kerma rate.

(d) Put the dosimeter below the plates, positioned towards the X ray tube in a central position. Make sure that this position does not affect the AERC control.
(e) Adjust the distance between the focal spot and the image receptor to match the clinical practice distance, aiming for 100 cm or the closest possible to it, as with other tests. Document the final SID used.
(f) Expand the collimators to cover the entire image receptor area.
(g) Determine and note the distances from the focal spot to the image receptor and from the focal spot to the radiation detector (instrument).
(h) Using a clinical imaging protocol, perform an exposure on the PMMA plate, and log the dosimeter reading, kV, mA and image receptor settings. Then, conduct this exposure twice more.
(i) Replicate the above step across a spectrum of clinically utilized FOVs, dose rate settings and imaging protocols typical in clinical settings, as well as with different attenuator thicknesses.

5.2.18.4. Analysis and interpretation

Adjust all measured values by applying the relevant correction factors, such as pressure, temperature and calibration coefficients for the instrument. Then, correct the measured ESAK rates to the reference point using the inverse square law. During the assessment, remember that ESAK accounts for backscatter, whereas the incident air kerma rate does not. In angiography machines, the patient entrance reference point is determined by the manufacturer and primarily varies with the type of interventional system, as specified in the relevant IEC standard [3, 32, 44].

Note: Instead of taking direct measurements, the displayed air kerma rates can be used, provided that appropriate corrections are applied.

5.2.18.5. Baselines and tolerances

The manufacturer of the equipment often provides a reference statement on incident air kerma rate or national regulations specify certain values for ESAK. If there are no national regulations, then the following values can be used from various international standards or guidelines.

For fluoroscopy modes [3]:

(a) With 20 cm PMMA:
  (i) Normal mode ≤50 mGy/min;
  (ii) High dose rate ≤100 mGy/min.
(b) For any thickness of patient, with PMMA:
  (i) Normal mode ≤100 mGy/min;
  (ii) High dose rate ≤200 mGy/min.

For cine or acquisition modes with 20 cm PMMA [3]:

(a) Variation ≤20% with respect to the baseline values;
(b) 0.03–0.12 mGy/image for interventional cardiology.

5.2.18.6. Frequency

Annually.

5.2.18.7. Corrective actions

If tolerances are surpassed, first reset the experimental set-up before conducting another test. Any failure to meet these standards needs to be approached with caution, as it might indicate issues with either the equipment or the protocol, or possibly both. Such situations necessitate expert analysis and often lead to further inquiry. Should this investigation pinpoint equipment faults, it is imperative to reach out to the maintenance service or the X ray system supplier. If it reveals imaging protocols problems, they ought to be addressed within other areas in the QAP [3, 40, 44].

5.2.19. Maximum air kerma rate

5.2.19.1. Description and objective

The objective of this test is to evaluate the maximum air kerma rates of fluoroscopy equipment. Radiation output levels are generally restricted by local or national regulations, and this test can be adjusted to ensure compliance with those regulations.

5.2.19.2. Equipment

An appropriate radiation measurement device that measures X ray air kerma rate. Either a solid state detector or ionization chamber can be used if it is calibrated for the X ray beam spectra being measured.

5.2.19.3. Procedure

(a) Centre the measuring instrument within the X ray field free-in-air (i.e. no table or mattress pad in the X ray beam path).

(b) Ideally, the detector needs to be placed within the plane where the air kerma rate is to be measured. If this is not possible, the measurement must be corrected to that plane. For C-arm fluoroscopy systems, the maximum air kerma rate is determined at 30 cm from the face of the image receptor assembly, regardless of the SID.

(c) If the fluoroscopy equipment has variable SID, this test must be conducted at both the minimum and maximum SID to ensure that the output accurately adjusts with the changing SID.

(d) Collimate the X ray field as small as reasonably achievable while ensuring that the measuring device is completely irradiated.

(e) Place sufficient attenuating medium in the X ray beam to drive the output to the maximum rate. Too much attenuation may terminate the exposure on some systems, while too little attenuation may not drive the system to the maximum output. The attenuator(s) need to be placed at least 10 cm from the measuring device toward the image receptor.

Note: Be sure that the attenuating medium fully covers the useful portion of the X ray image so that no bare, unattenuated X ray beam exposes the image receptor.

(f) Record the technique parameters (kVp and mA), ensuring they are at the system's maximum settings, along with the measured air kerma rates.

(g) Maximum air kerma rate measurements should be taken for various fluoroscopic system settings:
   (i) For all fluoroscopic dose curves (there are typically 2–3);
   (ii) For the largest and smallest FOV;
   (iii) Highest pulse rate (or continuous if system offers this option) and at least one other clinically used pulse rate.

(h) Repeat the test using a commonly utilized clinical FOV and cine frame rate for at least one clinically used acquisition.

5.2.19.4. Analysis and interpretation

The air kerma rates need to be recorded for each of the conditions listed above. It is useful to also compare them to measurements from previous evaluations.

5.2.19.5. Baselines and tolerances

Local or national regulations may set tolerances for this test; if so, then those values supersede any provided in this publication. Otherwise, for free-in-air measurements, under narrow-beam geometry, in the appropriate plane, with a calibrated dosimeter, the tolerances are as follows:

Fluoroscopy normal air kerma rate limit = 88 mGy/min;
Fluoroscopy high air kerma rate limit = 176 mGy/min.

Acquisition mode currently has no applicable air kerma rate limit. Note: air kerma rates measured under this geometry may exceed 2000–3000 Gy/min. Appropriate thickness and use of a Cu spectral filter will help keeping the air kerma rates at lower levels. In any case, caution needs to be used when this cannot be avoided and consider lowering the maximum achievable rates.

5.2.19.6. Frequency

Acceptance test, annually thereafter, and following major repair or replacement of the X ray tube, image receptor or software change that would affect the AERC logic.

5.2.19.7. Corrective actions

Corrective actions are similar to those of Section 5.2.18.7. Review the exposure parameters, calculations and the configuration of the set-up. Redo the test, fully reconfiguring the geometrical arrangement. Based on the deviations observed, it might be required to restrict or halt operations with the machine until repairs are completed. Conduct a comprehensive analysis to identify potential sources of error or variation in the results.

5.2.20. Table and table pad attenuation evaluation

5.2.20.1. Description and objective

The objective of this test is to evaluate the attenuation properties of the X ray procedure table and patient mattress. These objects are within the projected X ray beam for most gantry angles, and their attenuation ought to be evaluated. The accompanying documents for the imaging system ideally should specify the maximum value of the attenuation equivalence, according to the IEC 60601-2-42 standards.

5.2.20.2. Equipment

An appropriate radiation measurement device that measures X ray air kerma or air kerma rate. Either a solid state detector or ionization chamber can be used if it is calibrated for the X ray beam spectra being measured.

5.2.20.3. Procedure

(a) Using a posterior-anterior projection (0 degrees of primary and secondary gantry rotation), set the procedure table height to approximately isocentre.
(b) Set the X ray beam field size in the plane of the measurement to be approximately 10 cm × 10 cm.
(c) Centre the measuring instrument within the X ray field free-in-air (i.e. no table or mattress pad in the X ray beam path). Extend the measurement device off the end of the procedure table, if possible.
(d) If a radiographic exposure at a fixed technique is possible (preferable method):
    (i) Protect the image receptor with lead or other appropriate attenuators.
    (ii) Make a radiographic exposure at ~100 kVp and ~1 mAs; record the air kerma.

(iii) Place the measuring device on top of the procedure table and mattress and move the table to recentre the measuring device within the field, keeping the source to chamber distance and the X ray field size the same as for the previous measurement.

(iv) Make another radiographic exposure at the same exact settings as for the previous exposure; record the incident air kerma.

(v) Optional: Take a series of measurements in service mode to determine an attenuation curve as a function of beam quality.

(e) If a radiographic exposure cannot be used, perform the same steps as in step (d), but do so in the fluoroscopic mode and measure the air kerma rate instead of the integrated air kerma.

(f) If fixed technique operation of the system is unavailable, measurements can be acquired under the table, free-in-air and then above the table. The measurements ought to be corrected to the same plane by the inverse square correction.

5.2.20.4. Analysis and interpretation

The ratio of the air kerma (or air kerma rate) measurements with the table and pad in the beam to that of the free-in-air measurement will yield the transmission factor (the fraction of the incident X ray beam transmitted through the table and mattress).

$$\text{Transmission factor} = \frac{K_{\text{with tab and pad}}}{K_{\text{free-in-air}}} \tag{17}$$

5.2.20.5. Baselines and tolerances

Typically, the transmission should be between 0.6 and 0.8 at 100 kVp. If the transmission factor at 100 kVp falls below 0.6, the attenuation properties of the table and mattress need to be examined. Only tables and mattresses designed for use in X ray procedures should be utilized, as others may excessively attenuate the X ray beam.

5.2.20.6. Frequency

During acceptance testing and following replacement of the procedure table or mattress.

5.2.20.7. Corrective actions

If the transmission factor is too low, this will unduly burden the X ray tube, and the patient dose will be overestimated by the dose indices.

5.2.20.8. Clinical considerations

The X ray table and mattress attenuation needs to be considered in the calculation of patient peak skin dose estimates and for effective dose considerations for procedures or projections where the table and mattress are in the primary beam path.

5.2.21. RDSR and DICOM image data verification

5.2.21.1. Description and objective

DICOM RDSR and/or DICOM images are used to assess patient dose. DICOM dose report needs to contain relevant data for estimating DRLs or estimating patient dose using commercial or open source software. DICOM RDSR and DICOM images need to be checked for accuracy. RDSR needs to be checked for completeness: some tags such as patient height and weight or dimensions of X ray field are non-mandatory [61] and are sometimes not completed automatically, which may hinder dose calculations. Other tags are proprietary to the manufacturers and not readable by the user[2]. As a result, a CQMP responsible for different brands of equipment ideally should be able to interpret dose reports and RDSRs that are structured differently, making sure the information is properly used for calculation of peak skin dose or skin dose mapping.

5.2.21.2. Equipment

Any phantom can be used for this purpose.

5.2.21.3. Procedure

PACS and/or DACS configurations are required in order to configure the modality to send the RDSR to an appropriate destination, capable of receiving,

[2] As an example, for one vendor, the content of the procedure description tags has now been replaced by the term 'PROC DESC,' which does not allow the type of procedure to be identified.

displaying and archiving the reports. Note that not all PACS systems can accept DICOM RDSRs, while other systems may accept the RDSRs but not be capable of displaying the reports.

Expose the phantom using the most used fluoroscopy and cine mode(s). Make several X ray events while recording the technical and geometric parameters. Finalize the procedure and send the images and RDSR to the PACS and/or the DACS, if they are not automatically configured.

5.2.21.4. Analysis and interpretation

Retrieve the phantom image from PACS or DACS. Verify the consistency and accuracy of the image DICOM header and RDSR. Compare technical and geometric information with the data recorded earlier.

Verify the RDSR in DICOM format to ensure it is complete.

5.2.21.5. Baselines and tolerances

DICOM headers and values in RDSR need to match exactly.

All mandatory fields in RDSR ought to be complete and accurate; however, be aware that the table position is sometimes incorrectly recorded.

5.2.21.6. Frequency

Acceptance testing and following any software upgrade that may affect the DICOM RDSR.

5.2.21.7. Corrective actions

If any discrepancy is put forward, initiate engineering service.

5.2.22. Cone beam CT image quality and radiation dose evaluation

Both the European Federation of Organisations for Medical Physics (EFOMP)-IAEA-European Society for Therapeutic Radiology and Oncology (ESTRO) joint working group and AAPM have recently released the QC protocols for CBCT. The EFOMP-IAEA-ESTRO protocol [62] is centred on image quality and radiation output. In contrast, the EURADOS WG12 in Annex 6 [63] offers a literature review on radiation dose. The AAPM report provides a comprehensive review of image acquisition and reconstruction characteristics of C-arm CBCT systems and gives guidance on QC for C-arm systems with volumetric imaging capability [64].

On the basis of these reports, image quality is assessed primarily from phantom images. No image quality criteria are defined for clinical images, as this technology is probably too recent. Dose measurements in CBCT are challenging and come with several limitations, such as the size and positioning of the phantom, as well as the non-uniform distribution of the beam, etc. Radiation output reproducibility should ideally be measured using one of the various available methods, such as the CT dose index, P_{KA}, image receptor dose or exposure index.

This publication does not intend to introduce a new method for evaluating CBCT image quality and radiation dose, as there is currently no established consensus on this matter within the medical physics community.

6. OPTIMIZATION OF IMAGE QUALITY AND RADIATION DOSE

6.1. INTRODUCTION

FGIPs are essential for diagnosis and treatment, but at the same time they expose patients and healthcare professionals to ionizing radiation, which can lead to adverse health effects. Optimizing the image quality and radiation dose is critical to ensuring that patients receive the necessary diagnostic or therapeutic benefits while the associated risks are minimized.

6.2. GENERAL PRINCIPLES OF RADIATION PROTECTION

The guiding principle of radiation protection of patients, which is particularly important in FGIP, is 'as low as reasonably achievable. This principle means avoiding exposure to radiation that does not have a direct benefit, even if the dose is small.

The International Commission on Radiological Protection (ICRP) established a system of radiation protection based on three fundamental principles: justification, optimization and individual dose limitation. ICRP report 103, published in 2007 [65], provided updated guidance on these fundamental principles to promote appropriate radiological protection. ICRP has also published several publications related to radiation protection in imaging [66–69]. The principle of justification states that any decision that affects the radiation exposure situation ideally should bring more benefits than harm [34,

35]. The principle of optimization indicates that the likelihood of exposure, the number of people exposed, and their individual doses ought to be kept as low as reasonably achievable, considering economic and societal factors. Finally, the principle of application of dose limits is related to the individual and applies in planned exposure situations, where the total dose to any person from regulated sources ideally should not exceed the appropriate limits stated by ICRP and relates to the staff working within the FGIP departments and to the general population. This Section is specifically dedicated to exploring the aspect of optimization. It operates under the presumption that the necessary criteria for justification have already been satisfactorily met. This approach ensures a focused and thorough methodology for optimizing radiation exposure, while simultaneously acknowledging the fundamental importance of the justification process within the overall framework of radiological protection.

6.3. GENERAL PRINCIPLES OF OPTIMIZATION

To achieve an optimized medical imaging procedure, all the factors and quantities that affect image quality need to be adjusted to provide a clear visualization of anatomical structures and signs of pathology or injury while balancing the associated radiation risks. Optimizing FGIPs is a challenging task that cannot be achieved through a straightforward or simple process. Having satisfactory clinical image quality and reducing radiation exposure at the same time requires more than just altering technical parameters or following simple steps. The complexity of modern machines, the rapid pace of technology development and the availability of various post-processing options and software algorithms, along with the use of sophisticated exam protocols, add to the complexity of the process.

Although manufacturers have introduced radiation protection tools and committed to dose reduction, the process still requires careful consideration of many factors. Another important factor is that after X ray equipment installation, the manufacturer typically provides training focused on the basic operation of the system rather than on utilizing tools to achieve good quality images with minimal radiation dose. Additionally, when the equipment is used in clinical areas outside of radiology departments, such as surgical theatres, personnel may not have the time or opportunity to fully comprehend the available software features or other system tools. Finally, image processing software has advanced to a point where it has become the most important aspect, affecting both radiation dose and image quality. This ever expanding variable has multiple and conflicting effects on image quality and on potential procedure and patient safety risks.

Collaboration among various professionals is essential for successful optimization. This involves three main activities: (1) assessing and enhancing the performance of equipment; (2) customizing exam protocols; and (3) addressing staff behaviour. Manufacturers ideally should not only provide more radiation dose optimization tools but also train users on how to apply pre-set examination protocols. The presence of a CQMP that actively participates and leads the optimization process is particularly important.

6.4. ASSESSING IMAGE QUALITY

The initial step in enhancing clinical image quality involves evaluating the produced images, focusing on three key aspects: contrast, detail and noise. While contrast can be modified during image viewing through windowing, detail and noise levels are generally established during image capture and processing. Collaborative efforts are made to determine if the detail and noise levels are appropriate or need to be adjusted (refer to Section 6.6.1 for further information). Understanding the image quality needed for specific imaging tasks allows the team to embark on a continuous quality improvement journey, harmonizing image quality with other factors like radiation dose management and imaging speed.

Beyond technical assessments, a vital component of this optimization process is gathering direct feedback from radiological medical practitioners ideally after a procedure. This real time feedback is crucial, as it reflects the practical utility of the images in a clinical context. Radiological medical practitioners, having just completed the procedure, can provide valuable insights into the adequacy of image quality for diagnosis or intervention. Their perspectives on whether the image detail, contrast and noise levels were sufficient to carry out the procedure effectively are instrumental in guiding quality improvements. By incorporating this immediate feedback loop into the 'continuing quality improvement' process, the optimization team can more effectively balance image quality with other critical factors such as radiation dose management and image acquisition time. This approach ensures that the images not only adhere to technical quality standards but also fully support the clinical needs and expectations of the radiological medical practitioners.

6.5. PHYSICAL AND CLINICAL FACTORS AFFECTING IMAGE QUALITY AND DOSE

There are numerous factors that affect the quality of the displayed image and the patient radiation dose. Ideally the best image quality with regard to high contrast resolution and low contrast visibility ought to be obtained at the lowest possible dose. This goal can be achieved only at the price of a compromise that is driven by the level of image quality necessary for the clinical task. Several factors to consider and their anticipated effects are described below. Most of these parameters require service level access to view or change within a given imaging protocol. All FGIP equipment has image protocol databases that contain numerous X ray technical parameters, image processing parameters and image display parameters. Many of those parameters are proprietary to the vendors and unavailable or unknown to the end user. However, there are common parameters that clinical teams need to be aware of and may optimize with vendor or service level support. Always work with the imaging equipment vendor and/or image quality specialists when making changes in order to avert any unintended consequences and confirm all changes made with phantom measurements first before clinical implementation. Moreover, radiological medical practitioners also know which image quality levels are required for their specific clinical tasks (insert and guide a catheter, dilate an occluded vessel, see an implanted device, etc.), although they do not use the same terminology as CQMPs to express image quality.

6.5.1. Technical considerations

Technical considerations include:

(a) Image receptor target dose:
  - (i) All protocols need to have discrete target image receptor doses set for each mode of imaging within a given protocol. Most FGIP systems will use SNR target doses, while some newer systems may also use CNR target doses or beam spectra.
  - (ii) Typically, these image receptor doses are expressed as a set nGy/pulse and can potentially range 3 orders of magnitude. More specifically:
    - — In fluoroscopy mode, values may be as low as 5 nGy/pulse and range to upwards of 100 nGy/pulse.
    - — In acquisition mode, values may range from less than 100 nGy/pulse to upwards of 5000 nGy/pulse.
  - (iii) There is obviously a direct relationship between this setting and the image quality and radiation dose:
    - — High contrast resolution (also known as limiting spatial resolution): This will not be affected by this setting, assuming that the resolution is not limited due to excess noise in the image.

— Low contrast visibility: The contrast will generally improve with increasing target doses, until the SNR is sufficiently high that quantum noise is no longer apparent.

— Patient dose: With all other parameters held constant, a linear relationship between the target image receptor dose and the $K_{a,r}$ or P_{KA} would be expected.

(b) Default pulse and frame rates:

(i) Typically, the default rates and pulse length can be configured differently for each protocol in the protocol settings.
Note: For some vendors the rates can be changed from their default values by the end user, while other vendors do not allow the default pulse rate to be changed.

(ii) This parameter will directly affect the temporal resolution and radiation output of the fluoroscopy equipment, so care needs to be taken to set this parameter appropriately for the clinical task, for each imaging protocol and for each dose curve. Even for systems that do allow the user to change the pulse rate, it is often the case that the default is predominately used.

(iii) High contrast resolution: This will not be affected by this setting as long as the resolution is not limited due to excess noise in the image or excessive recursive filtration.

(iv) Low contrast visibility: The contrast will generally improve with increasing pulse rates, but this will depend on the amount of recursive filtration applied.

(v) Patient dose: With all other parameters held constant, including dose per pulse, there is a linear relationship between the pulse rate and the expected air kerma rates.

(c) Recursive filtration (also known as temporal filtration, k-factor):

(i) All interventional imaging systems use recursive filtration in some manner, but how and to what extent varies from vendor to vendor and even potentially software version to software version of the same manufacturer and model.

(ii) Essentially, recursive filtration is an image formation process that weights and sums a set of image frames over a short period of time. The intended benefit is an increase in the SNR of the displayed image by weighting and summing several acquired image frames.

(iii) If the recursive filtration is set to integrate a large number of images, then more image frames will be used in the summing process. This will increase the SNR, thereby reducing the appearance of noise in the image for the same image acquisition settings. However, objects in motion will be blurred out and lose contrast, since they are not in the same location within the image from one frame to the next.

(iv) High contrast resolution: This will not be affected unless there is no motion in the image, in which case the resolution will be reduced for the areas under motion due to blurring.

(v) Low contrast visibility: This will generally increase with increasing amounts of recursive filtration since there is an expected gain in SNR from summing multiple frames. However, any motion will degrade the contrast of the object in motion, since the object will be in different locations from one image frame to the next within the group of images being integrated.

(vi) Patient dose: Recursive filtration is an image processing feature and will not in itself affect the radiation output. However, greater amounts of recursive filtration will increase SNR, and, in this way, either recursive filtration can be used to increase image quality (reduce noise) at the same radiation dose, or, for the same original SNR, the radiation dose can be decreased.

(d) Beam quality: Typically, vendors will have target values for the factors that determine the X ray beam production. These values provide only a target, or starting point, as the fluoroscopy equipment AERC will vary the settings with respect to the overall target image receptor dose on the basis of the set SNR or CNR.

(i) kV – The X ray tube voltage potential is the primary factor affecting the X ray beam quality and could range from ~50–125.

— High contrast resolution: To a first approximation, kVp will not affect the resolution. However, low kVp settings (in conjunction with other factors) may require the system to change focal spot size, and that will impact the resolution.

— Low contrast visibility: The X ray beam quality, largely determined by the kV, substantially affects low contrast resolution in the image. The subject contrast (inherent contrast within the subject being imaged) is directly dependent on the beam quality.

— Patient dose: The relationship of kVp and dose is complicated, especially for AERC systems. If all other parameters are held constant (manual exposure), increasing kVp will increase the dose as more X rays with greater average energy will be produced. However, this is rarely, if ever, the scenario for FGIP equipment, where AERC is used to maintain a constant dose to the image receptor. In that case, increasing kVp will lower the dose with all other parameters held constant, but care must be taken with regard to the meaning of 'dose'. Skin entrance absorbed dose will decrease, whereas exit absorbed dose will remain the same (to a first approximation).

(ii) mA – The X ray current is defined differently by the manufacturers. Some use a time averaged value, while others use a peak instantaneous value.

— High contrast resolution: To a first approximation, mA will not affect resolution, as long as the resolution was not initially noise limited. However, high mA settings may require the fluoroscopy equipment to change focal spots size and that will impact the resolution.

— Low contrast visibility: Although mA will not affect the subject contrast, it does play a role in the overall image contrast. The product of mA and s is the mAs per pulse and directly affects the image noise magnitude, and therefore the CNR.

— Patient dose: There is a direct linear relationship of mAs (product of mA and time for each pulse width) and radiation dose, given all other parameters held constant.

(iii) ms – The X ray pulse width is the beam on time in milliseconds for each X ray pulse, for pulsed X ray imaging.

— High contrast resolution: To a first approximation, ms will not affect resolution, if there is no motion occurring within the pulse width. For imaging performed where there is considerable motion (e.g. cardiac), short pulse widths are required to reduce intra-frame blur.

— Low contrast visibility: Although ms will not affect the subject contrast, it does play a role in the overall image contrast. The product of mA and ms is the mAs per pulse and directly affects image noise magnitude and therefore the CNR.

— Patient dose: There is a direct linear relationship between mAs and radiation dose, given all other all other parameters held constant.

(iv) Cu filter: Spectral filters can be used for two primary purposes. At low kVp, they are used to generate a narrow beam energy with relatively high effective energy. At high radiation doses, the filters can be used to increase the X ray beam quality, thereby reducing the skin entrance dose.

— High contrast resolution: To a first approximation, Cu filtration will not affect the resolution. However, high amounts of Cu filtration may require the use of a larger X ray focal spot, which will affect the resolution.

— Low contrast visibility: Cu filters will harden the X ray beam, with all other parameters held constant. Therefore, high amounts of Cu filtration may reduce contrast.
— Patient dose: Additional Cu filtration will reduce the skin entrance absorbed dose, with all other parameters held constant.

6.5.2. Clinical considerations

For each unique clinical procedure, there are specific image quality requirements and radiation dose considerations that likely differ and will not be consistent even within a procedure. In most cases, the equipment's automatic exposure control takes care of this compromise. The operator's role will then be to select the most appropriate dose saving protocol suitable for the clinical circumstances. For example, the use of a 'low dose' protocol may lead to an increased level of image noise but may still provide acceptable image quality for diagnosis. For example, Fig. 46 shows a lesion of the ostium of the diagonal which is identifiable using a 'low dose' protocol. It needs to be noted that dose rates that are too low can result in images that are not clinically useful. In such cases, the radiological medical practitioner may consider using cine to overcome poor fluoro quality. Thus, a continuous balance between image quality and radiation dose is crucial. The crux of this balance is not just to reduce radiation dose but to do so without compromising the diagnostic or therapeutic utility of the image.

The clinical image quality levels can be described through three degrees of visibility of the anatomical structures represented on the radiological image, as described below:

(a) Level 1 — Visualization: the organs and structures are detectable in the volume of investigation.
(b) Level 2 — Critical reproduction: details of anatomical structures are visible but not necessarily clearly defined.
(c) Level 3 — Visually sharp reproduction: anatomical details are clearly defined.

As mentioned in Section 6.4, gathering direct feedback from radiological medical practitioners, ideally immediately after a procedure and using but not limited to the above degrees of visibility of anatomical structures, is an important part of a process to ensure that the images produced meet the practical needs of clinical procedures.

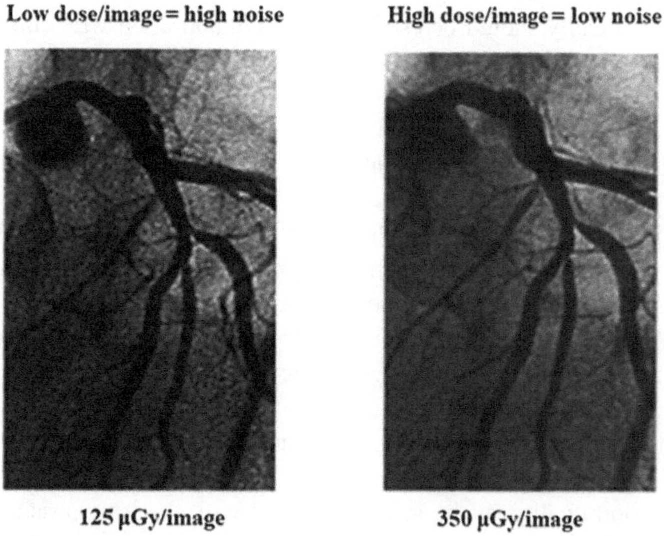

FIG. 46. Example of two image qualities acquired at different dose rates. The lesion of the ostium of the diagonal is identifiable at the low dose rate.

6.6. THE OPTIMIZATION PROCESS

6.6.1. Optimization team

To ensure efficient and successful optimization of medical imaging process, an optimization team ideally should be established. This team can consist of various professionals and include at least an FGIP radiological medical practitioner, a CQMP, an MRT and a vendor image quality specialist. Each member of the team has a unique role in the process of optimization [70, 71]. The FGIP radiological medical practitioner is responsible for providing feedback on maintaining sufficient task specific image quality. The CQMP guides the optimization process and optimizes the examination protocols while ensuring sufficient image quality at an appropriate radiation exposure. The MRT ensures that the optimized exam protocol is feasible in the clinical workflow and executed correctly. The vendor image quality specialist is necessary to coordinate potential changes in the image processing settings, since adequate descriptions of the related settings are typically not provided or available from the vendor.

The team collaborates to decide which clinical tasks, examination types and X ray modalities require optimization. As the frequency and types of examinations can differ greatly across institutions, this decision is best taken at an institutional level. Once the exam types are decided, clinically appropriate image quality requirements ideally should be established in collaboration with clinicians. This ensures that optimization is carried out in a comprehensive manner that considers the specific clinical needs of the institution.

6.6.2. QAP as part of optimization

Section 4 outlines in detail the terms of QAP, QA and QC. Establishing a QAP is crucial for successful optimization. A well-defined QAP ensures that all processes and outcomes meet the highest standards of quality and efficiency. It plays a pivotal role in identifying areas for improvement, setting benchmarks for performance and providing a systematic approach to achieving these goals. By rigorously monitoring and evaluating each step of a process, a QAP helps in pinpointing inefficiencies, reducing errors and ensuring consistency. According to the WHO, a QAP is an organized effort to consistently produce diagnostic images of good quality, at the lowest cost and with the least possible radiation exposure to the patient [31]. The optimization process requires striking a balance between image quality and patient dose, while also considering clinical factors such as the use of contrast agents. Quality assurance and dosimetry in diagnostic radiology in general have been extensively studied by international organizations such as IAEA and the European Commission [44, 72–75], as well as medical physics societies or organizations, such as IPEM or AAPM [42, 47, 76]. Regardless of which protocol or guidance is followed, establishing a QAP is crucial for effective optimization.

6.6.3. Establish baseline image quality and radiation dose levels

Once the optimization team decides which examinations or protocols require optimization, the next step is to establish a baseline of patient image quality and radiation exposure levels for those specific examinations. Ideally this is done by performing appropriate dose measurements concurrently with image quality assessment. Image quality assessment is performed using specific criteria decided in advance. Technical image quality assessment typically involves contrast, noise and spatial resolution measurements. Clinical image quality evaluation is done by the FGIP radiological medical practitioner. Once compiled, dose index statistics are compared with local, typical, regional or national DRLs, or with other relevant benchmark values, taking into account clinical image quality adequacy [77, 78].

DRLs are considered an important tool for optimization. However, misinterpretation of DRL use can occur, such as the assumption that if patient doses are below DRL, this means that adequate optimization has been achieved. Therefore, the optimization team ideally should consider factors such as clinical acceptability [79]. Caution is required not to apply DRL values to local institutional datasets that are inconsistent with the underlying factors used to establish the reference DRLs (e.g. dissimilar procedure complexity, different patient populations (race, age, gender, BMI, etc.), differences in equipment capabilities, etc.) [80, 81].

6.6.4. Modification of technical parameters

The process of modifying exam protocols is a complex task that requires the expertise of a CQMP in collaboration with the other members of the optimization team. With the advancement of medical technology, various tools are available to achieve the optimization goal, but it ought only to be attempted in collaboration with the vendor image quality specialist with a deep understanding of the machine's performance and how each technical parameter, post-processing algorithm or other features can affect image quality and radiation dose. The CQMP will need to carefully study all technical documentation to comprehend each machine's characteristics and optimization tools. Additionally, the vendor image quality specialists can provide advanced knowledge, best practices, and tips and tricks for the specific modality in question.

Several publications have addressed radiation protection concerns in FGIP but very few have dealt with the topic of image quality as well. The literature emphasizes the need for a combined approach of equipment protocol modification and operator behaviour as far as dose optimization is concerned, but none of them comments on the level of image quality. Instead of solely modifying protocol characteristics, various strategies have been suggested, such as digital processing, customized exam protocols, removal of antiscatter grid in specific clinical cases (e.g. spinal DSA), reduction of pulse rate and frame rate, default use of the fluoroscopy mode at low level, and continuous dose monitoring and patient follow up. While new technology features of modern X ray angiography machines hold promise for dose optimization, their use and impact on image quality and dose needs to be fully understood [82]. Therefore, a comprehensive evaluation of X ray machines during commissioning and regular quality control by a CQMP is crucial.

6.6.5. Suggested framework for optimization process

With input from the full optimization team, define the clinical requirements of the procedure and the necessary image quality to achieve successful clinical results.

(a) Special considerations ought to be given to:
   (i) Paediatric patients: Specific protocols need be developed, with the possibility of multiple ranged protocols based on age, weight, BMI or other metrics.
   (ii) Pregnant patients: Specific imaging protocols may be necessary or specific clinical processes developed to address the pregnant patient's needs.
   (iii) Obese patients: Specific imaging protocols ought to be developed to ensure sufficient ranges of image quality and dose rates to ensure clinical efficacy.
(b) Note that 'copying' imaging protocols from existing equipment, even for the same vendor and model of equipment, is not advisable. There are many differences in the protocol settings for these systems that differ from one software version to another and certainly between models or vendors.
(c) The clinical image quality requirements may differ for each clinical procedure, and even within the same procedure during different procedure steps. However, common image quality considerations include the following:
   (i) High contrast resolution requirements (which vessels, structures or devices need to be visualized).
   (ii) Low contrast visibility requirements (resolvability of low subject contrast objects).
   (iii) Temporal resolution requirements and constraints (available pulse rates, pulse widths, extent of recursive filtration, etc.).
(d) Define appropriate starting protocol settings, ensuring that low dose techniques are used to start the procedures while also ensuring a wide range of available image quality and dose rates to ensure clinical efficacy.
(e) After protocol change, but prior to patient use, reacquire phantom data to ensure changes resulted in the expected outcomes and no unintended consequences.
(f) Evaluate the protocol as a team during clinical procedures. The protocol ought to be used for a variety of patient procedures with a range of procedure complexity and image quality demands (various ages, heart rate, BMI, etc.) to ensure protocol adequacy over a full range of patients.
(g) Iterate the steps above until the image quality needs are met at the lowest reasonable clinical radiation dose.

6.7. METHODOLOGY FOR CLINICAL OPTIMIZATION

Optimization requires the appropriate selection and configuration of a complex set of design features for the fluoroscopy system, tailored to the clinical tasks and required level of image quality. It is of utmost importance to put all efforts of optimization into those examinations that require the highest levels of radiation dose, namely cardiac, neuro and vascular applications.

Most optimization efforts have been made in the cardiac domain (Appendix IV). Technical strategies for lowering radiation dose while maintaining an acceptable image quality that is fit for the clinical purpose are described below (Section 6.7.1). Manufacturers are constantly developing more efficient detectors and processing methods, and suppliers are producing tools that allow increasingly more complex procedures increasingly quickly. Clinical efforts are still needed when preparing the procedures in order to make sure the quickest approach is always chosen. The easiest way to prove that an optimization effort has been successful is by recording local and/or typical DRLs before and after the optimization process and comparing them while checking the success rate of the procedures. The most efficient way to compare DRLs is to establish them by clinical indication (see Appendix IV) to normalize as much as possible the dosimetric and imaging aspects. This consequently allows a true comparison between operators or types of equipment and allows the optimization process to be performed.

6.7.1. Optimization for paediatric patients

6.7.1.1. General suggestions

Paediatric interventional imaging encompasses a broad spectrum of unique clinical complexities and body types. As with all fluoroscopic imaging protocols, it is essential to establish specific protocols tailored to the requirements of the clinical procedure, ensuring sufficient image quality without unnecessary radiation exposure. Interventional procedures on paediatric patients need to be conducted using only appropriate fluoroscopy systems equipped with AERC, radiation dose monitoring and tools that save patient dose such as LIH, virtual collimation and virtual positioning. Moreover, certain considerations unique to paediatric patients are often not encountered or deemed necessary in adult procedures.

One of the primary considerations is the potential for excessive motion, which can be challenging to avoid in young paediatric patients. This motion may be due to greater involuntary anatomical movement, such as a higher cardiac rate, or difficulties in controlling voluntary movement. As a result, several settings may need optimization for paediatric patients compared to adults, including the following:

(a) Lower levels of recursive filtration ideally should be utilized to maintain spatial resolution, temporal resolution and contrast by limiting image frame integration. This might lead to slightly higher image noise, but a higher dose per pulse can be employed if the noise level becomes excessively high.
(b) To accurately image the targeted anatomy, higher fluoroscopic pulse rates and acquisition frame rates might be required. Since the radiation dose rates scale with fluoroscopic pulse and acquisition frame rates, any increases in these rates ought to be implemented judiciously and only to the extent necessary.
(c) Reducing the pulse width to the smallest possible extent can help lower the dose per pulse for a given mA, providing a more precise image by controlling the amount of motion integrated within a specific pulse or frame. This also reduces the effect of motion on intraframe blurring.
(d) Collimation, while beneficial in all cases, ideally should be maximized in paediatric imaging to limit superfluous radiation to organs outside of the clinically significant volume. This is particularly crucial in very young patients, whose organs are considerably smaller than those of adults. Small reductions in the X ray beam area can substantially decrease the radiation dose to organs peripheral to the anatomy of interest, specifically excluding the patient's eyes, thyroid, breasts and gonads whenever possible.
(e) High levels of spectral filtration, often upwards of 0.9 mm of Cu, are typically used. For small paediatric patients, the reduced attenuation due to lower tissue volumes results in lower kVp. In such instances, increased amounts of spectral filtration can be used without substantially degrading contrast, simultaneously hardening the X ray beam for better transmission and thereby reducing the overall radiation dose.
(f) Gridless imaging can significantly reduce the patient's radiation dose and may be considered for smaller paediatric patients, since there is less tissue volume being imaged and therefore fewer scatter photons generated. However, for larger patients, this may result in degraded image quality, specifically reduced contrast due to increased noise levels.

6.7.2. Optimization for pregnant patients

6.7.2.1. General suggestions

Medical exposure of pregnant patients requires a strict process of justification, in which benefits and risks to both patient and the conceptus (embryo and foetus) ideally should be taken into consideration. Non-ionizing imaging modalities such as magnetic resonance imaging or ultrasound ideally should be preferred if fit for the clinical purpose.

Pregnant patients might inadvertently be exposed to radiation either early in pregnancy or when the pregnancy is not known. Under such circumstances, calculating the dose received by the conceptus might be necessary. There are web-based tools designed for estimating the conceptus dose from diagnostic and interventional X ray procedures. Imaging facilities need to have clear notices alerting potentially pregnant patients about the risks of X ray exposure to the conceptus. It is crucial to ascertain the pregnancy status of all patients of childbearing age before conducting X ray examinations that could expose the conceptus, using a standardized inquiry form. FGIPs like endovascular coiling in trauma, vascular dissection or malformation bleeding, stents or nephrostomy placements, radiofrequency ablations or ERCP ideally should be optimized, especially those that include the conceptus in the direct beam.

There are, of course, cases where a pregnant patient may need to undergo an FGIP, and there are specific scenarios in which undergoing an FGIP is not only medically necessary but also the most appropriate course of action for a pregnant patient, despite the inherent potential risks associated with radiation exposure. However, it is crucial to acknowledge and address the fears and concerns that might arise for the patient and her family in these situations. Pregnant patients, often concerned about the wellbeing of their unborn child, might experience significant anxiety about the impact of radiation exposure. This apprehension can sometimes lead to reluctance or even refusal to undergo medically necessary procedures. It is important to engage in open and empathetic communication with patients and their families, providing clear, accurate information about the risks, benefits and safety measures in place for FGIPs during pregnancy. Educating patients about how these procedures are carefully managed to minimize radiation exposure to the foetus and discussing the potential health implications of avoiding necessary medical interventions are key aspects of this conversation.

6.7.2.2. Establish baseline image quality and radiation dose levels

The health and safety of both pregnant patients and their developing foetuses are of utmost priority in medical imaging procedures. It is imperative to establish baseline image quality and radiation dose levels specifically tailored for this vulnerable patient group. Protocols need to be in place for different gestational ages and clinical indications. Technical parameters such as kVp, mA, pulse rate and filtration need to be adjusted to provide optimal imaging for the specific clinical need, taking into account the increased body thickness and altered anatomy of pregnant women.

6.7.2.3. *Optimization strategy*

Practical ways to control conceptus dose during FGIPs are proposed below:

(a) Consider using ultrasound guidance for catheter insertion and choose a route that reduces conceptus dose.
(b) Choose the equipment that delivers the smaller dose with the most appropriate protocol for the pregnant patients. A special protocol for that purpose needs to be developed in each department in collaboration with the CQMP.
(c) Precisely collimate the beam to limit exposure.
(d) Minimize the exposure duration to reduce radiation dose.
(e) Utilize the highest feasible tube potential to optimize image quality and minimize dose.
(f) Limit the use of magnification mode to avoid unnecessary radiation increase.
(g) Position the X ray tube as far from the patient as practicable while keeping the detector near to minimize dose.
(h) Opt for low dose rate and pulsed fluoroscopy to decrease radiation exposure.
(i) Utilize the LIH or video loop feature to reduce the need for additional exposures.
(j) Minimize the radiation dose during DSA.
(k) Assess the optimal condition of the maternal bladder for the required projections to enhance image quality and minimize exposure.

6.7.2.4. *Evaluation of image quality and radiation dose*

The highest image quality ought to be determined that will achieve the highest possible standard while using the lowest dose of radiation necessary to meet the clinical objectives of the examination. While the importance of optimizing image quality and radiation dose for pregnant patients in FGIP is understood, current literature remains notably sparse on this specific topic. The unique physiological changes and sensitivities that occur during pregnancy necessitate a more dedicated exploration into the ideal balance between image quality and radiation dose. More comprehensive studies are urgently required to establish robust guidelines and best practices tailored for this patient group. Such research will not only address a significant gap in the existing literature but will also pave the way for safer and more effective imaging protocols for pregnant patient, ensuring the wellbeing of both patient and foetus.

7. CONCLUSIONS

This publication places a significant emphasis on a comprehensive QAP within an FGIP facility and encompasses elements such as performance testing, meticulous assessment of clinical images and thorough optimization process. It serves as a valuable guide, shedding light on the critical aspects to consider when setting up an FGIP facility, exploring various designs of FGIP systems and their integral components while offering insights into the most recent technological advancements. The publication delivers intricate steps regarding performance testing, highlighting the most common pitfalls and errors that could potentially compromise the results or the interpretation of specific tests. In relation to the optimization process, it elaborates on the multifaceted factors that can directly affect image quality and radiation exposure. To this end, it also provides practical clinical examples that demonstrate how to effectively apply optimization in daily practice.

The task of optimizing FGIPs is a daunting and complex challenge, one that does not lend itself to an overly simplistic or direct process. Achieving an optimal balance between good quality clinical images and reduction of radiation exposure goes beyond merely adjusting technical parameters or adhering to a basic procedural guideline. It is a process fraught with intricacies due to the complexity of modern machinery, the swift progress in technological advancements, and a myriad of post-processing alternatives and software algorithms at user's disposal. Furthermore, the introduction of increasingly sophisticated examination protocols further escalates the complexity of the task at hand. The drafting committee agreed to provide a proposed framework for the optimization process, accompanied by clinical examples. The present guide demonstrates that the task of optimization, despite its complexity, is not only necessary but also achievable. This feasibility, however, relies heavily on understanding of the processes involved and the ability to implement them diligently. It requires fully comprehending the intricacies and nuances of each step, its implications and outcomes, and the connections between these steps.

Appendix I

FGIP EXAMPLES

The following compendium details a diverse array of procedures categorized into three primary parts: (a) vascular imaging and diagnosis, (b) vascular interventions and (c) non-vascular interventions and invasive diagnostic procedures. The list is not exhaustive and serves as an educational tool and reference for various professionals in the medical field.

(a) Vascular imaging and diagnosis:

- (i) Angiographic diagnosis (with or without treatment) with catheters and guide wires;
- (ii) Selective catheterization techniques;
- (iii) Normal and variant arterial anatomy;
- (iv) Normal and variant vascular anatomy of systemic veins;
- (v) Normal and variant vascular anatomy of the portal venous system;
- (vi) Normal and variant pulmonary vascular anatomy;
- (vii) Normal and variant anatomy of the lymphatics;
- (viii) Arterial vascular disorders;
- (ix) Systemic venous disorders;
- (x) Dialysis access disorders;
- (xi) Portal venous disorders;
- (xii) Pulmonary arterial disorders;
- (xiii) Lymphatic disorders.

(b) Vascular interventions:

(i) Application of principles of vascular interventions and planning.
(ii) Peripheral arterial intervention;
(iii) Treatment of arteriovenous malformation;
(iv) Aortic interventions;
(v) Gastrointestinal bleeding, arterial;
(vi) Mesenteric ischemia;
(vii) Visceral aneurysm interventions;
(viii) Renal artery interventions;
(ix) Hepatic arterial interventions;
(x) Arterial trauma interventions;
(xi) Uterine artery interventions;
(xii) Bronchial artery embolization;
(xiii) Transjugular intrahepatic portosystemic shunt;
(xiv) Variceal obliteration, sclerosis;
(xv) Balloon occluded retrograde transvenous obliteration of gastric varices;
(xvi) Portal vein dilation and stenting;
(xvii) Portal vein recanalization;
(xviii) Pulmonary arteriovenous malformation;
(xix) Management of pulmonary embolisms, catheter interventions;
(xx) Reproductive tract interventions;
(xxi) Upper extremity/superior vena cava occlusive disease;
(xxii) Lower extremity venous occlusive disease;
(xxiii) Foreign body retrieval;
(xxiv) Vena cava filter placement;
(xxv) Transvenous biopsy;
(xxvi) Venous sampling for endocrine disease;
(xxvii) Haemodialysis access management;
(xxviii) Venous access;
(xxix) Lymphangiography;
(xxx) Thoracic duct ablation.

(c) Non-vascular interventions and invasive diagnostic procedures:

(i) Lumbar puncture/spinal tap;
(ii) Pancreatic inflammatory disease, stenting or drainage;
(iii) Enterocutaneous fistula;
(iv) Follow-up care and procedures;
(v) Placement of tunnelled peritoneal drain;
(vi) Lymphocele drainage and sclerosis, seroma drainage and sclerosis;
(vii) Oesophageal intervention;
(viii) Gastrostomy and gastrojejunostomy;
(ix) Percutaneous cecostomy;
(x) Percutaneous jejunostomy;
(xi) Gastrointestinal tract stenting (duodenal and colonic);
(xii) Percutaneous transhepatic cholangiography;
(xiii) Biliary drainage;
(xiv) Biliary dilation and stent;
(xv) Percutaneous cholecystostomy;
(xvi) Percutaneous nephrostomy;
(xvii) Nephroureteral dilation and stenting;
(xviii) Renal cyst sclerosis;
(xix) Suprapubic cystostomy;
(xx) Hysterosalpingography and fallopian tube interventions;
(xxi) Treatment of uterine leiomyomata (fibroids);
(xxii) Ablation of liver masses;
(xxiii) Chemical ablation of liver masses;
(xxiv) Tunnelled pleural drainage catheters;
(xxv) Airway dilation and stenting;
(xxvi) Percutaneous vertebroplasty;
(xxvii) Thermal ablation of bone lesions;
(xxviii) Vertebral height restoration;
(xxix) Percutaneous disc interventions;
(xxx) Selective nerve root block;
(xxxi) Stellate ganglion block;
(xxxii) Facet injections.

Appendix II

UNCERTAINTIES

Given the complex nature and significant implications of understanding and managing uncertainties in scientific measurements and calculations within FGIPs, it is crucial to allocate a dedicated section to this topic in the guidance. This section is vital, as it addresses the diverse sources of uncertainty that affect the accuracy and reliability of results. It is important to acknowledge that every scientific measurement and calculation inherently involves some level of uncertainty. Within the framework of the methods and procedures outlined in the preceding sections, these uncertainties arise from various sources and significantly influence the outcomes. This appendix offers a detailed examination of these uncertainties, providing a structured analysis as per below:

(a) Every measuring device, regardless of its complexity, has inherent inaccuracies due to factors like slight manufacturing variations, wear and tear, or external influences. Additionally, the calibration process, which aims to enhance accuracy, can also introduce small errors. The uncertainties from the device's inaccuracy and its calibration are generally small but notable. Regular calibration and maintenance can help reduce these discrepancies.

(b) The set-up process for any measurement can be prone to potential errors. Factors such as positioning or alignment can lead to deviations. These set-up inaccuracies typically contribute a small portion to the total uncertainty. Proper training and strict adherence to procedures can help minimize these uncertainties.

(c) Measures taken to ensure the accuracy and reliability of measurements inherently bring their own uncertainties. Depending on the specific procedure, these can range from 5% to as high as 30%. For instance, higher observed uncertainties, such as 25% for the P_{KA} and 30% for the air kerma, necessitate the application of correction factors to accommodate these deviations and ensure more accurate results.

On another note, patient skin dose is very important in FGIP because values can often reach the threshold of deterministic effects. When assessing the uncertainties associated with calculating the maximum skin dose, it is crucial to understand that this dose metric can incorporate uncertainties from previously mentioned sources [74]. Often, this originates even from dose metrics taken from the X ray equipment. As a result, the cumulative uncertainty can be significant, potentially reaching values as high as 40%. This underscores the importance of factoring in these uncertainties during interpretation and application.

Manufacturers often base their published uncertainties on the results from direct, intrusive testing protocols, which involve closely interacting with the system or device under test. These tests are typically conducted under controlled and ideal conditions to ensure precision and reliability. These invasive tests are designed to directly measure parameters within the equipment and offer high accuracy. On the other hand, in practical CQMP scenarios, testing often relies on non-invasive tools, which, while safer and more feasible for routine use, can introduce additional layers of uncertainty. These additional uncertainties can be substantial enough to cause such tests to fail, even when the equipment is functioning within acceptable manufacturing tolerance levels. This is because non-invasive tools may not detect specific details or the particular conditions in which the manufacturer's direct, intrusive tests were performed. Therefore, it is important to factor in these additional uncertainties when interpreting CQMP test results, with the understanding that deviations might stem from the testing methodology itself rather than from actual equipment performance issues.

Appendix III

PATIENT DOSE ASSESSMENT

For complex FGIPs, the potential risk to the patient's skin due to radiation exposure is a paramount concern. Additionally, it is essential to acknowledge the clinical variability in how different patients may react to a given skin dose. Individual factors such as age, skin type, pre-existing conditions and overall health can influence the skin's response to radiation exposure. Therefore, the same calculated skin dose may result in varying degrees of tissue reaction or injury across different patients. This variability necessitates a cautious and individualized approach when assessing potential risks and communicating with patients about the implications of their skin dose during and after fluoroscopic procedures.

Given this risk of skin injury, it becomes vital to continuously monitor cumulated dose quantities throughout the procedure. Once the procedure concludes, these quantities need to be meticulously recorded [83]. To ensure patient safety and prompt intervention, specific trigger levels ideally should be established in advance. These levels serve as critical benchmarks, indicating when patient follow up is necessary and when proactive measures ought to be taken to manage potential tissue reactions. Furthermore, it is essential to adopt a holistic perspective when assessing radiation exposure. This means considering not just the radiation dose from the current procedure but also accounting for radiation exposure from past procedures and the likelihood of future procedures [2, 67, 70]. The most appropriate trigger level at the time of publishing this guidance is 5 Gy of displayed $K_{a,r}$ [84]. In situations where the equipment in use does not provide a display of $K_{a,r}$, a value of P_{KA} needs to be chosen as a trigger value. For instance, in cardiac applications, 300 Gy.cm^2 can be considered appropriate, whereas for vascular applications, a value of 500 Gy.cm^2 can be more suitable based on typical field sizes. Once the trigger value has been exceeded, either in a single procedure or over multiple procedures, the PSD or a skin dose map needs to be calculated. Fortunately, the modern technological landscape offers various software tools designed for this purpose. These tools can facilitate real time (online) PSD calculations during the procedure or allow for retrospective (offline) calculations once the procedure is concluded. It is worth noting that, for repeat procedures, real time calculations are often not feasible, making offline evaluations the go-to solution. In cardiac procedures, where the fluoroscopic beam frequently shifts its position, the recorded air kerma often overshoots the actual PSD. Conversely, in procedures where the beam remains relatively stationary, air kerma levels can be significantly lower than the true PSD. This discrepancy can lead to trigger levels that are overly optimistic, highlighting the importance of understanding the nuances of each procedure and adjusting protocols accordingly.

Appendix IV

EXAMPLES OF OPTIMIZATION PROCESS: CARDIOLOGY

Interventional cardiology FGIP encompasses a multitude of factors that require careful consideration. Key among these is the specific clinical indication that necessitates such a procedure, a vital component in the optimization process. Neglecting this aspect can render optimization efforts less relevant, as operators would then apply a broad approach to diverse interventional procedures, disregarding the specific technical complexities and medical issues that drive these procedures. Once an FGIP is determined based on clinical indication, the complexity of the procedure needs to be considered. For instance, applying a one-size-fits-all approach to cardiac angioplasty without differentiating between monovascular and trivascular lesions oversimplifies the whole process. At the same time, ensuring acceptable image quality that addresses the clinical problem is crucial. Insufficient image quality could lead to suboptimal outcomes, as it may compromise the effective execution and success of the FGIP. The following paragraphs present a detailed overview of image quality assessment methodologies in cardiology, accompanied by pragmatic guidelines for implementing optimization strategies in interventional cardiology.

IV.1. IMAGE QUALITY EVALUATION

For an effective anatomical assessment of the vascular lumen, which is critical for formulating a relevant diagnosis and determining potential therapeutic strategies, sufficient quality of the radiological image is indispensable. Given that radiological images are inherently imperfect representations of the actual imaged object, establishing clear criteria for evaluating these images is crucial. Some of these criteria can be assessed directly by the operator through visual examination of the information displayed on the monitor, while others necessitate more sophisticated technical or physical tools. It is essential to achieve a harmonious balance between objective measurements and subjective perceptions, as image quality and patient radiation dose are intrinsically linked to this equilibrium.

IV.1.1. Objective evaluation method

Physical and quantitative, this method allows the assessment of the intrinsic characteristics of the image thanks to the use of physical and/or electronic test objects, which ensure reproducible and comparable results from one imaging system to another. This method requires specific skills and resources that cannot be easily implemented on a daily basis. It is routinely applied as part of the QC tests described in Section 5.

IV.1.2. Subjective evaluation method

This method, which is qualitative and reliant on the visual perception and training of the operator, facilitates the estimation of the image's intrinsic characteristics. It focuses solely on the inherent properties of the clinical image itself and is used in routine clinical daily practice. Certain quality criteria are suggested in the following paragraphs, which can be considered a reference framework that considers the following aspects:

(a) The level of image quality useful to establish a diagnosis or to conduct a therapeutic procedure is neither consensually nor regulatory defined.
(b) The subjective assessment method is more easily applicable in the everyday clinical practice than the objective method and it allows specifying the level of clinical image quality required and is sufficient for addressing the specific clinical problem.
(c) The goal of optimization is to achieve sufficient image quality to address the clinical problem at the lowest possible dose.

The application of the subjective method necessitates the consideration of three key factors: clinical, technical and dosimetric.

A. Clinical criteria

The clinical criteria express the degrees of visibility of the anatomical structures represented on the radiological image. Three levels of visibility are proposed, and an example is given in Fig. 47.

(a) Visualization: The organs and structures within the examined volume are identifiable.
(b) Critical reproduction: Structures relevant to the specific clinical indication are discernible to the degree necessary for accurate diagnosis. This includes the terms:
    (i) Reproduction: Anatomical structures are visible, though not always with clear definition.
    (ii) Visually sharp reproduction: anatomical details are distinctly defined.

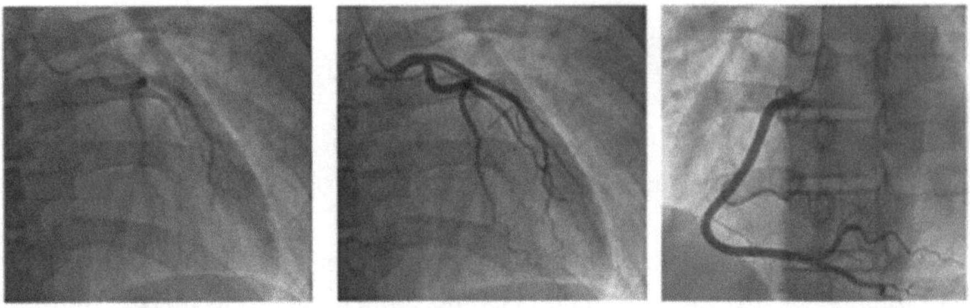

Visualization Reproduction Visually sharp reproduction

FIG. 47. Presence of patient's arms in the primary beam; absence of wedge filter.

B. Technical criteria

Table 3 shows some suggested technical criteria that can be referred to in the daily practice of diagnostic coronarography procedures.

Figure 48 provides an example of an inappropriate technique. Superimposing the patient's arms produces an additional attenuation of the primary beam by an unwarranted increase in the thickness traversed, resulting in a significant increase in the kV and mA exposure parameters, poorer image quality and, most importantly, a significantly higher dose delivered to the patient. To overcome this problem, a wedge filter could have been used in this case. This would clearly have improved the clinical image quality by improving the contrast of the image and proper use of AERC.

C. Dosimetric criteria

The dosimetric criteria are the DRL, K_{PA} and/or air kerma baseline values determined for cardiology FGIPs. If a procedure is more complex than usual (e.g. treating a bifurcation lesion or dealing with chronic total occlusions), it may require more fluoroscopy time, additional angiographic runs or higher image quality settings, thus justifying a higher dose. Additionally, larger patients or those with challenging anatomy may require higher doses to achieve adequate image quality. In less resourced clinical settings, operators occasionally need to use older or less advanced X ray equipment that may not have the latest dose reduction technologies, leading to higher baseline dose levels than more resourced settings.

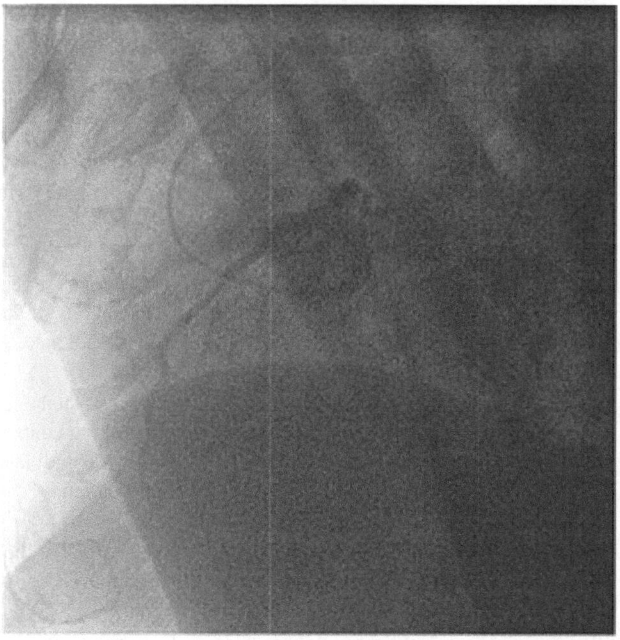

FIG. 48. Presence of patient's arms in the primary beam; absence of wedge filter.

IV.2. OPTIMIZATION IN THE EVERYDAY CLINICAL PRACTICE

Certain practical tips that can significantly enhance the optimization process in interventional cardiology to ensure improved outcomes and patient safety are explored below.

IV.2.1. Arterial access

Femoral and radial access require special technical training in both puncture and catheter handling. Due to their different and variable positioning (right versus left) in relation to the source of the scattered radiation (patient), they may lead to variations in radiation exposure for both the patient and the operator. Recent studies have shown that the radiation dose levels are similar for both femoral and radial access routes. Additionally, patients report significantly lower levels of discomfort and pain with the use of radial access [85, 86].

TABLE 3. CARDIAC IMAGING TECHNICAL CRITERIA

Technical criteria	Interest	Comments
Field of View	Full coverage of cardiac area. Avoid travelling during acquisition.	Spatial resolution does not depend on FOV (for FPD).
Complete filling of coronary artery (contrast medium)	Analyse the actual lumen at least to the first flow-limiting lesion (90–95%).	Image contrast depends on the homogeneity of the filling by the contrast medium.
Length of run	Visualize collaterals and the downstream bed (including if artery occluded).	Images at the beginning of the run without contrast medium (except if studying coronary calcifications).
Positioning the arms outside the field	Avoid humeral overlays. Limit the dose of brachial skin.	Skin lesions of the arms reported.
Number of projections	Identify the three segments of each coronary artery and their dividing branches >1.5 mm.	Each segment of the coronary artery ought to be visualized through two orthogonal views, if possible.
LAO 90° projection	Visualize the LAD downstream bed. Clear the cross of the heart.	Myocardial bypass. Preoperative bypass surgery.
Rotational angiography	Visualize the coronary network from multiple views. Decrease the amount of contrast agent.	Requires an automated injector. Iso centrism required.

IV.2.2. Catheterization and contrast agent

During catheterization, various technical decisions, such as the choice of catheter diameter and type, directly influence the filling of intracoronary vessels with a contrast agent. Furthermore, the presence of the contrast agent enhances the absorption of X rays, consequently increasing the radiation dose to the heart.

Achieving clear visualization of the arterial lumen is strongly linked to the effectiveness of this contrast filling. Below are some suggestions:

(a) Catheter selection optimization: The appropriate catheter diameter and type ideally should be chosen on the basis of the specific clinical scenario. Adequate vessel opacification needs to be ensured, even if smaller diameter catheters, which reduce contrast volume, are used.

(b) Volume and concentration of contrast agent: The volume and concentration of the contrast agent ought to be carefully managed. Adequate arterial lumen visualization needs to be achieved using the minimum necessary volume of contrast to reduce X ray absorption and, consequently, the radiation dose to the heart. The total amount of contrast used ideally should be monitored, particularly in patients with conditions that could be exacerbated by high volumes of contrast media. The quality of arterial lumen visualization needs to be continuously assessed during the procedure, and adjustments to the use of the contrast agent ought to be made accordingly. Repeat injections ought to be avoided whenever possible.

IV.2.3. Injector device

The timing of the contrast agent injection ought to be synchronized effectively with the imaging sequence, and an appropriate rate of injection needs to be used for uniform vessel opacification. Rheological aspects related to contrast agent viscosity, catheter size, cardiac flow rate, coronary size and pathology (e.g. occlusion) will determine the quality of coronary filling. In addition, the use of an automated injector makes it possible to significantly improve staff radiation protection (working further away from the patient).

IV.2.4. Use of X ray equipment

(a) Operation mode

Contemporary angiography systems are equipped with a variety of operational protocols, each offering different levels of dose rates and image quality. These dose rates can range from ‘low’ to ‘medium’ and then to ‘high’ in both fluoroscopy and cine modes. The lower dose settings are paired with significantly high X ray beam filtration to minimize skin dose. Operators ideally should become thoroughly familiar with all the functionalities of the X ray equipment to utilize it most effectively.

(b) Tube–detector–patient geometry

The tube to detector distance during angulations of the C-arm is dictated by the thickness of the patient. The patient to detector distance ought to be as small as possible. Recent technical advances by manufacturers allow for automatic adjustment of this distance. The distance between the patient and the X ray tube needs to be as large as possible. The most favourable relative positioning of the tube–detector–patient assembly needs to be maintained throughout the procedure to minimize overexposure of the skin. In the vast majority of cases, these different geometry settings remain highly dependent on the operator's vigilance and compliance with the information displayed on the control monitor.

(c) Wedge filter

The wedge filters are attenuating lamellae placed at the exit of the X ray tube and whose movements can be either automatically inserted in the X ray field according to the X ray beam projection or manually controlled by the operator. Their thickness decreases towards the centre of the image so that they cause partial attenuation of the primary beam. The positive effects associated with the rational use of wedge filters are a decrease in the intensity of X rays reaching tissues with low attenuation, such as the lungs, thereby reducing the dose to these areas as well as significantly improving image contrast through optimal choice of exposure parameters.

(d) Beam collimation

Collimation of the primary X ray beam is essential in the optimization process. A motorized set of leaded shutters at the exit of the X ray tube allows the size of the primary beam to be adapted to the field of anatomical region of interest. Several positive effects are associated with the proper use of beam collimation for both the patient and the operator:

— An improvement in the intrinsic image quality by increasing the visual contrast of the image (reduction of the surrounding noise level);
— A reduction of the projected area of the primary beam on the patient's skin;
— A superimposition risk reduction of the different beams and thus a reduction of the maximum peak skin dose in case of multiple X ray projections;
— A reduction in the volume of exposed tissue and therefore in the dose delivered to the organs;

— A reduction in the amount of scattered radiation reaching staff.

(e) Magnification

The visualization of fine anatomical details is possible with magnification. The physical size of the detector, expressed in cm, determines the maximum FOV available. Changing the FOV (for example, a field of 20 cm diagonal to a field of 15 cm) leads to an enlargement of the image through projecting a central part of the detector onto the same display surface (monitor screen size). The magnified image has improved spatial resolution that is visually detectable by the operator, even though the intrinsic resolution of the detector is not improved because it is physically limited by the pixel size, but with increased radiation dose delivered to the patient.

(f) Practical tips

Specifically for coronary angiography, a relatively standard protocol is applied by the operator based on the number and types of C-arm projections according to coronary artery network and criteria for visualizing the anatomical structures to be identified. Some practical advice is given in Table 4 to avoid recurrent drawbacks.

Percutaneous coronary intervention is more complex compared with coronary angiography. Procedures involving multiple or more challenging vessels typically require longer fluoroscopy times and potentially more angiographic runs. Additionally, the number and length of stents implanted play an important role. Longer or multiple stents demand more precise and extended manipulation, which in turn can lead to increased demands for image quality and radiation exposure. Bifurcation lesions or chronic total occlusions often necessitate additional imaging and intervention time. Understanding and managing these variables are key to optimizing patient outcomes while minimizing radiation exposure, which underscores the need for tailored approaches in each percutaneous coronary intervention [87].

Table 4. PRACTICAL OPTIMIZATION TIPS FOR CORONARY angiography

Recurrent problems	Practical tip for optimization
Motion blurring	— Increase frame rate (3.75 to 7.5 fps), ideally with smaller pulse width.
Visual contrast insufficient (diaphragm or chest)	— Use a wedge filter.
Images without intra-coronary contrast medium	— Limit exposure before injection. — Automated injector (rotational angiography).
Too many images per run	— Limit to 50 frames per run on average. — Default frame rate set at 3.75 fps.
Travelling	— Use the fluoroscopy mode to locate the endpoint. — Travelling allowed for by-pass only (reduction of contrast medium).
Ventriculography: 30 fps	— Reduce frame rate to 15 or 7.5 fps, if possible.

Appendix V

EXAMPLE OF OPTIMIZATION PROCESS: CARDIAC ELECTROPHYSIOLOGY — LEADLESS PACEMAKER

Often when discussing approaches to fluoroscopy protocol optimization, the goal is radiation dose reduction. However, there are times when clinical demands require the best possible image quality to avoid harmful outcomes or to ensure a proper diagnosis. In these types of cases, the clinical demands justify increasing the radiation dose rates to provide the best possible image quality. What follows is a clinical example of such a clinical procedure. This is just one instance illustrating the potential need for increased radiation output and image quality to achieve the necessary clinical outcomes.

The leadless cardiac pacemaker represents a significant advancement in heart pacemaker technology, eliminating the need for wired leads. This small, self-contained device is directly implanted into the right ventricle of the heart. Roughly the size of a large vitamin capsule, it sends electrical impulses directly to the heart muscle to regulate its rhythm, paralleling the function of conventional pacemakers. Notably, leadless pacemakers eliminate the need for a surgical chest pocket and the placement of vascular leads into the heart, thereby significantly reducing associated complications.

The insertion method for a leadless pacemaker is considerably less invasive than that of its traditional wired counterpart. It is introduced through a catheter that is navigated via a vein in the leg and through the inferior vena cava towards the right side of the heart. The device is then deployed in the apex of the right ventricle. Once positioned, it generates electrical signals to modulate the heartbeat, addressing instances where the heart rhythm is too slow or irregular. Thanks to its wireless and compact design, the leadless pacemaker reduces the risk of infection and lead-associated complications and offers a safer, more comfortable alternative for patients requiring pacemakers. The implantation of the leadless pacemaker into the cardiac muscle involves a percutaneous procedure. The process begins with the introduction of a catheter into the patient's femoral vein, located in the groin area. The pacemaker is loaded onto this catheter and then guided through the vein using fluoroscopic imaging. The device is optimally positioned within the right ventricle and then released from the catheter. Equipped with small, flexible tines, the pacemaker securely anchors into the heart tissue. Once the correct device positioning and functionality are verified, the catheter is removed from the body. To ensure the device functions properly, careful deployment is essential. The tines need to be securely embedded in the cardiac tissue to provide stable anchoring and a reliable electrical connection (Fig. 49). Given their small size and the cardiac motion, the tines are not easily visible during fluoroscopic imaging. Accurate placement is confirmed by tugging on the device to observe tine deflection, which indicates sufficient embedding. This critical stage of the procedure requires optimal image quality for clinician confidence. Accordingly, some institutions have implemented specific cine acquisitions at 30 fps, minimal recursive filtration, minimal additional copper spectral filtration, and a relatively high image receptor dose per pulse. This type of acquisition ought to be used selectively during the most crucial phase of the exam to provide the best possible image quality for clinicians when the demand for image quality justifies a higher radiation dose rate to the patient.

In the provided example, a nuanced scenario is encountered where the emphasis on minimizing radiation dose might inadvertently compromise clinical imaging quality. Such situations underscore the challenges faced by healthcare professionals in striking the right balance: they ideally should not only safeguard patients from excessive radiation but also ensure that the clarity and precision of the images obtained are suitable for accurate diagnosis and treatment planning. It becomes evident that the pursuit of reduced radiation ideally should not be at the expense of the clinical efficacy of the imaging. To navigate these complexities, it is crucial to embed robust training and educational programs. These programs need to emphasize the judicious use of imaging protocols and advocate for their deployment solely in situations where they are indispensable.

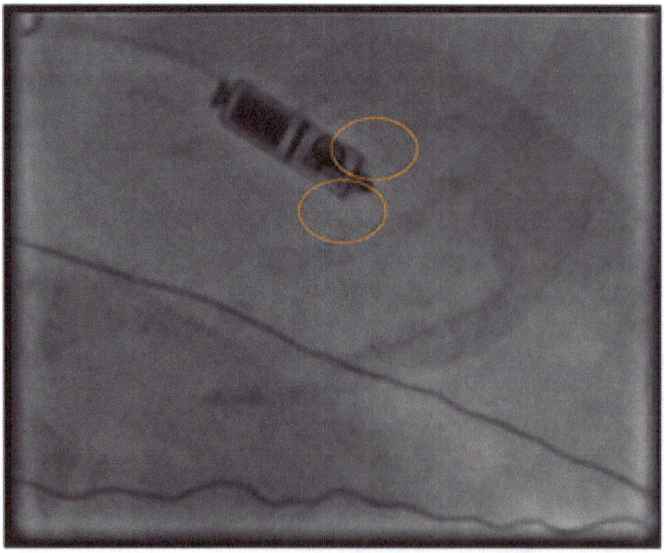

FIG. 49. X ray image of a leadless pacemaker with two of the embedding tines circled.

Appendix VI

EXAMPLE OF OPTIMIZATION PROCESS: ENDOSCOPIC RETROGRADE CHOLANGIOPANCREATOGRAPHY

VI.1. INTRODUCTION

Endoscopic retrograde cholangiopancreatography (ERCP) is a contrast-enhanced procedure of intra- and extrahepatic bile ducts, gallbladder and pancreas that uses a side-viewing endoscope to achieve selective cannulation of the bile or pancreatic duct. Since non-invasive cross-section examinations such as CT and magnetic resonance cholangiopancreatography provide good quality images of the hepatobiliary and pancreatic system facilitating diagnosis of a plethora of conditions, the vast majority of ERCPs are performed mostly for therapeutic purposes (stone extraction, stent placement, etc.). Moreover, the recently developed single operator cholangioscopy allows for a direct intraductal view of the ductal system with or without additional interventions (e.g. electrohydraulic lithotripsy, tissue acquisition, etc.) [88]. These dedicated cholangioscopes are advanced through the working channel of the duodenoscope and are increasingly integrated in the daily clinical practice.

VI.2. ESTABLISHMENT OF TYPICAL RADIATION DOSES IN ERCP

As previously mentioned in Section 2.5, these procedures can be performed in a diverse range of settings utilizing either an over-the-couch or under-the-couch X ray tube system, in an operating theatre, in the endoscopy unit with a mobile C-arm, or in an interventional imaging suite employing an angiography system. In the field of interventional endoscopy, there are not enough published data on DRLs, while reported ERCP DRLs depict a wide distribution even for the same procedure at the same facility. In a national multicentre study from Japan [89] that includes data from 11 162 ERCPs, the median or third quartile values of $K_{a,r}$ (mGy), P_{KA} (Gy.cm^2) and fluoroscopic time (min) were 69/145 mGy, 16/32 Gy.cm^2 and 11/20 min, respectively. On the other hand, in a Greek cohort of ERCPs, Tsapaki et al. [90] demonstrated that the DRLs for P_{KA} and fluoroscopic time, deriving from the third quartiles of the total sample, were 19 Gy.cm^2 and 8 min, respectively, while the correlation between P_{KA} and fluoroscopic time is efficiently described by a power equation.

There is evidence that the establishment and implementation of DRLs on a local or national level may contribute to the optimization. In a Finnish study by

Saukko et al. [91], the adaption of local DRLs led to lower mean values of dose area product (also called kerma area product or P_{KA}) and fluoroscopic time five years after their implementation compared to the era before their introduction.

The variance of reported P_{KA} values is also illustrated in guidance provided by international scientific societies. In the most recent European Society of Gastrointestinal Endoscopy guidelines published in 2012 [92], the mean values of P_{KA} reported for diagnostic and therapeutic ERCPs range from 3 to 115 Gy cm^2 and from 8 to 333 Gy.cm^2, respectively. Thus, establishment of DRLs in ERCP procedures as well as including fluoroscopic time as an indirect measure of radiation exposure or even more, adaption of more accurate quantitative indicators such as P_{KA} in the list quality metrics for ERCP could strongly contribute to optimization. Of note is that this is currently not included within the European Society of Gastrointestinal Endoscopy, American Society for Gastrointestinal Endoscopy or British Society of Gastroenterology or standards for ERCP, and none of the studies included subjective or objective image quality evaluation. On the other hand, there are multiple factors that may affect radiation doses during ERCPs including complexity of the procedure, patient-related factors, endoscopist experience, fluoroscopy time and differences among X ray systems. The abovementioned factors as well as clinical image quality, among other factors, ought to be taken into account when aiming to optimize ERCP procedure.

VI.3. FACTORS AFFECTING IMAGE QUALITY IN ERCP

VI.3.1. Patient related factors

The area of interest during ERCP procedures is the right upper abdominal quadrant for cholangiography and the central abdominal area for pancreatography. Historically, ERCP can be performed with the patient in various positions, namely the prone, the supine or the left lateral decubitus position. Appropriate selection of patient's position depends on endoscopist's preference, local expertise, need for endotracheal intubation as well as on patient comorbidities (e.g. respiratory pathology, obesity, spine and hip surgeries, etc.). Overall, the prone position (Fig. 50) is favoured during ERCP, since it is considered the most optimal one for successful papilla cannulation and for obtaining good quality fluoroscopic images. On the other hand, the supine position may improve visualization of hilar anatomy, facilitate airway management, and improve patient comfort. Improvement of the obtained fluoroscopic images (cholangiogram) can be achieved by adopting the fluoroscopy settings according to patient's body size

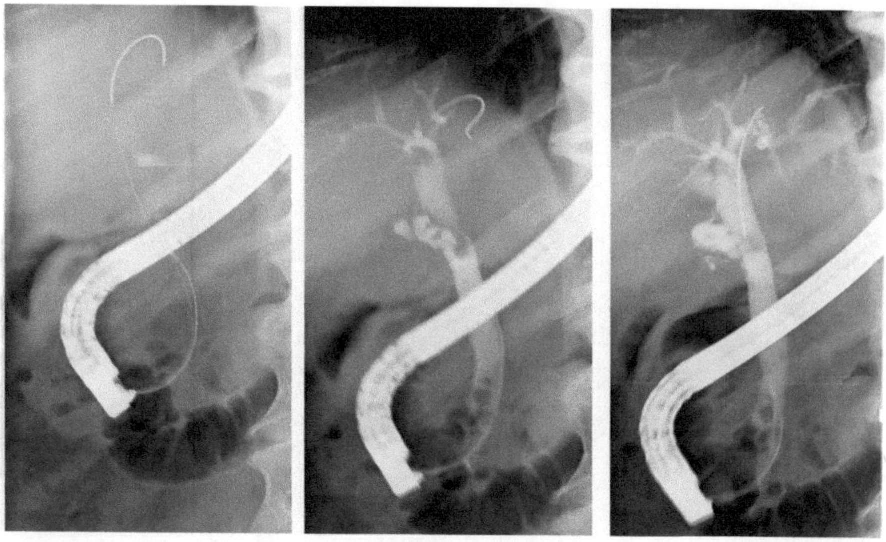

(a) (b) (c)

FIG. 50. ERCP performed with the patient in the prone position. (a) Guidewire assisted cannulation before contrast medium injection; (b) and (c) radiopaque deficits of 7–8 mm, compatible with small common bile duct stones and the remnant of cystic duct. The size of the common bile duct and the stones is estimated using the diameter of the duodenoscope as a reference.

and habitus. Finally, patient movement during the procedure may diminish the quality of image due to spine or arm projection.

VI.3.2. Procedure related factors

For ERCP procedures, obtaining a good quality cholangiogram (in the vast majority the common bile duct is the duct of interest) and the ability to interpret it in real time are the fundamentals for subsequent management (Figs 51 and 52). For successful optimization during ERCP several procedural factors ought to be considered. First, before initial exposure, the X ray tube needs to be positioned at the area of interest, while the X ray beam needs to be collimated to the smallest possible size for the area of interest. Second, pulsed instead of continuous fluoroscopy needs to be used, while at the same time the lowest acceptable image quality is selected by using magnification only when it is expected to significantly contribute to the ongoing task. Obviously, reverting to the lowest needed magnification ought to be the case once the higher resolution is no longer needed. Third, the direct anterior-posterior angle is preferable, although another

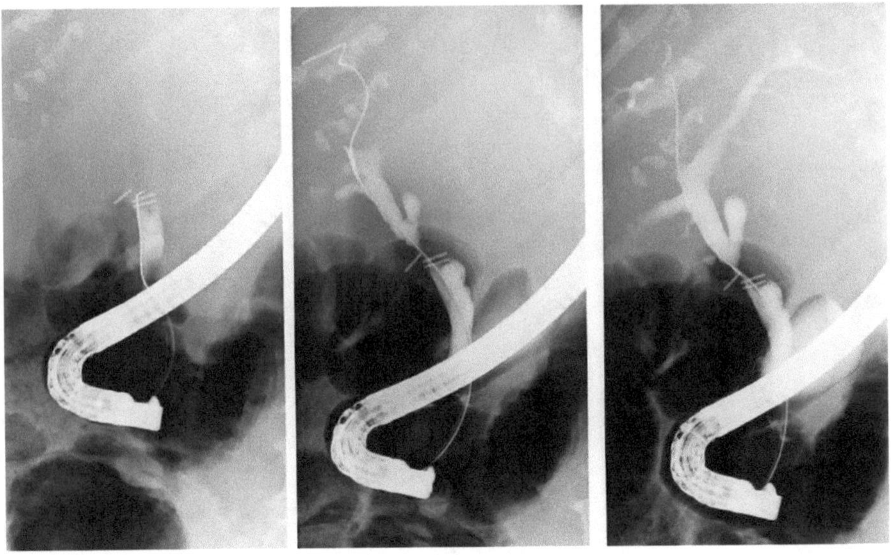

FIG. 51. Cholangiogram demonstrating a common bile duct stricture post cholecystectomy. Consecutive and meticulous image acquisition after injection of contrast medium starting from the distal common bile duct allows valuable information to be obtained, including the exact length and site of the stricture.

angulation may occasionally be useful, while hilar strictures might be preferably depicted in magnification mode, allowing for accurate guidewire manipulation. Fourth, it is of paramount importance to obtain a baseline image acquisition of the area of interest to be used as a baseline for comparison with subsequent fluoroscopic images taken after contrast injection. Skipping this step could cause confusion in the interpretation of radiology images due to calcifications in ribs or vessels, various artefacts from overlapping structures, or the presence of surgical or endoscopic clips or recently administrated contrast media. Moreover, it is important to keep in mind that the appropriate X ray field size is around 30–40 cm in order to capture images of the entire hepatobiliary tree, while discrepancies in the $K_{a,r}$ and P_{KA} values may suggest the use of a larger irradiation field in some procedures, a fact that could be attributed to non-collimated X ray field.

Finally, the 'fluoro save' feature of the X ray machine needs to be used whenever available. This obviates the need to acquire acquisition mode images only for archival purposes, thereby avoiding unnecessary patient radiation dose. For ERCP, the radiology as well as endoscopy images usually are part of the ERCP report. Early and delayed images are captured for photo documentation use. The ERCP X ray system ideally should be connected to the hospital PACS so that images are available for team review locally or remotely.

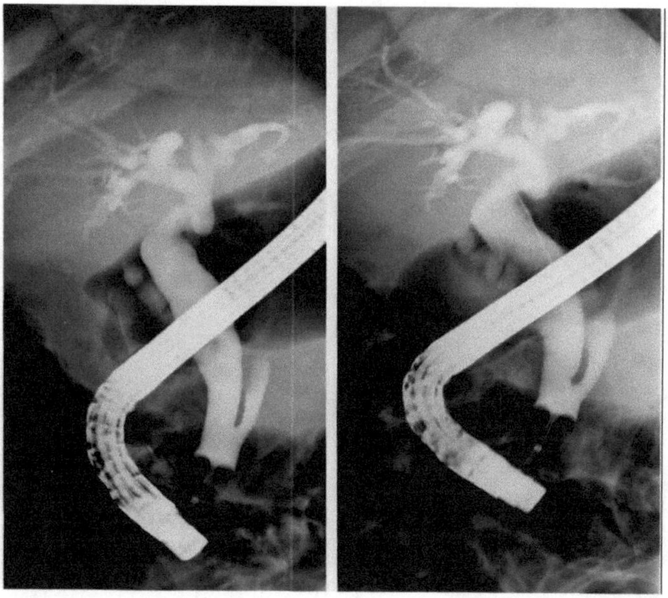

(a) (b)

FIG. 52. Balloon occlusion cholangiogram at the end of a procedure demonstrating (a) the absence of residual common bile duct stones and (b) low emergence of the cystic duct.

VI.3.3. Operator dependent factors

It is well known that optimizing image quality also depends on MRTs. Thus, it is of value that some members of the personnel are accordingly trained to assist in the ERCP procedures. In terms of controlling the footswitch, this is preferably carried out by the endoscopist compared to MRT in order to avoid unnecessary radiation exposure due to a delay in the process. Acquisition mode images significantly contribute to radiation exposure during ERCP.

The experience of the endoscopist and the ability to interpret fluoroscopy images are very important elements towards optimization [93]. It has been shown that ERCP performance by a low volume endoscopist was associated with a twofold increase in the patient radiation dose [94].

Another factor that could also lead to a reduction of the occupational dose is making the endoscopist aware in real time during the procedure of both the fluoroscopic time and the radiation doses, since this practice could potentially modify the endoscopist's behaviour (Hawthorn effect). Finally, preprocedural understanding of clinical pathology, anticipated anatomy based on preprocedural examinations such as MRI or CT and discussion of the expected therapeutic plan could proactively diminish the administrated amount of fluoroscopy.

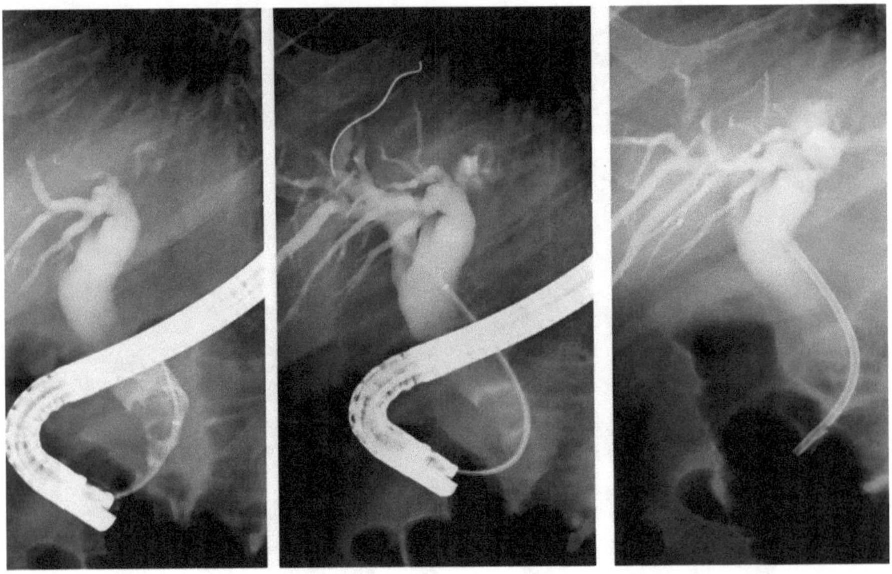

(a) (b) (c)

FIG. 53. This cholangiogrram demonstrates multiple radiopaque deficits compatible with common bile duct stones in the lower part of the common bile duct (a). Notable here is that the common bile duct is dilated (~15 mm) with a tapered appearance at its proximal part. Moreover, due to the dilation of the common bile duct, a significant amount of contrast medium had to be injected to acquire a good quality image of the duct (b). Finally, a temporary plastic stent was inserted (c).

VI.3.4. Underlying disease or indication for performing ERCP

Another factor that could affect the usage of radiation dose and image quality during ERCP is the procedure's complexity. The vast majority of ERCP procedures are performed to treat small (less than 1 cm) common bile duct stones or to place a stent in patients with subhilar strictures of the bile duct. According to the ERCP complexity grading system, these cases are graded as simple on the basis of anticipated technical difficulty and rates of success. On the other hand, more complex ERCP cases include the removal of large stones using mechanical lithotripsy, or intraductal lithotripsy using a cholangioscope as well as the treatment of proximal malignant biliary strictures. Complex ERCP cases have been associated with prolonged fluoroscopic time [95], $K_{a,r}$ and P_{KA} (Fig. 53).

VI.3.5. Equipment related factors

For the time being, there are no data directly comparing different types of X ray systems (over-the-couch or under-the-couch system, mobile C-arm, steady C-arm or angiography machines). Usually, the choice of the X ray machine that will be used for ERCP depends on factors such as the locally available equipment and rooms, expertise, hospital budget and multiuse procedural rooms versus dedicated FGIP rooms inside endoscopy units. However, in specialized referral endoscopy centres, dedicated ERCP rooms with X ray systems are permanently installed and allow for higher quality capabilities.

In everyday clinical practice, the ERCP rooms are more commonly equipped with mobile C-arm systems. These systems are characterized by their relatively small size and flexibility in moving together with a relatively lower cost compared to other systems. Moreover, data suggest that over couch X ray systems are not appropriate for ERCP due to increased scattered radiation [96]. Bowsher and Blott [97] showed that the radiation dose delivered to the body parts of the radiological medical practitioner above the table by an over-the-couch system is five- or sixfold higher than that delivered by an under-the-couch system, likely because of the increased radiation scatter above (instead of below) the patient in the latter case.

Over-the-couch systems expose the head of the radiological medical practitioner to a larger amount of radiation as measured by lens dose and fluoroscopy time in comparison with the under the couch systems [98]. This fact leads to measures such as the need to better protect the endoscopist's head and to modify the practice towards a more radiation friendly X ray system, wherever available.

Regarding the use of mobile C-arms, it is important to consider that appropriate beds are needed to accommodate the under-the-bed beam so as to focus in the area of interest. Ideally, both tables and mattresses provided by manufacturers should be appropriate for X ray imaging procedures.

Currently, there is a growing interest around ergonomic issues, and both the European Society of Gastrointestinal Endoscopy and the American Society for Gastrointestinal Endoscopy have published guidance for the correct set-up. There are data to support that the 32 inch size display monitor in endoscopy is preferable to 27 inches. On the other hand, there are no studies for the ideal size of fluoroscopy monitors or the preferable position inside the endoscopy suite.

Usually, the endoscopy equipment is attached to the endoscopy tower. The anaesthesiology team is often found near the patient's head. The position of the equipment (tower, displays, etc.) in the room varies. A large display monitor or a stack of display monitors are commonly situated next to the patient bed, while additional monitors are found in the control room. Evidence highlights the importance of ergonomics, particularly in positioning fluoroscopy and endoscopy display monitors close to each other, to enhance workflow efficiency and reduce fluoroscopy time.

VI.4. Establishment of baseline image quality criteria

There are no set quality criteria for obtaining a good cholangiogram in ERCP, and no studies have assessed the accuracy of cholangiogram interpretation during ERCP or interventions to improve this. In any case, the potential to optimize radiation usage while maintaining efficacy remains the key requirement.

The quality of images obtained during ERCP may be affected by the quality of fluoroscopy, the characteristics of contrast medium and the injection technique. Most commonly for ERCP, a high osmolality iodinated contrast medium is used, since its cost is much lower than low osmolality contrast medium and produces similar image quality with no significant differences in safety profile [99]. Undiluted full strength contrast medium is appropriate when fine imaging details are desired, such as in visualizing strictures and pancreatic duct anatomy, whereas half strength contrast medium (diluted 50-50 with saline) is preferable for assessment of choledocholithiasis in a dilated bile duct [95]. The ease of injection, especially through a small diameter catheter is greatly affected by contrast medium viscosity.

Factors contributing to reduced image quality level and possibly increased radiation dose to the patient include the following:

(a) Error in the protocol settings which establish the target kV and image receptor dose, etc.;
(b) Inappropriate distance between patient and X ray radiation source;
(c) Using continuous instead of pulsed fluoroscopy;
(d) Not using the collimators (worsens contrast);
(e) Not focusing solely to the field of interest;
(f) Not using the dilution and appropriate contrast material;
(g) Filling the ducts with too much contrast under pressure.

Ten simple steps to improve quality of cholangiogram images during ERCP include the following [99]:

(a) With the duodenoscope in front of the papilla, take a baseline image before injecting contrast medium into the common bile duct. This scout image helps to interpret findings that existed before contrast injection.

(b) Before cannulating the duct of interest, flush the catheter to completely remove the air and prevent air artefacts in the cholangiogram mimicking small stones.
(c) Take fluoroscopic sequencing images at early phases following contrast injection. Extensive filling of ducts may obscure findings such as small stones or the bifurcation view.
(d) Try gentle dynamic contrast injection to obtain better images. In stone cases, contrast injection needs to start from the upper common bile duct to prevent the transportation of small stones into the intrahepatic ducts where they can be more difficult to identify.
(e) Minimize contrast injection in selected situations such as in hilar strictures. Selectively cannulate with the guidewire and opacify only the ducts indented to drain with stent insertion.
(f) Use a guidewire to direct the sphincterotome and contrast injection into the selected duct. With the patient in prone or left decubitus position, the left hepatic ducts fill earlier than the right ones. Tilting the bed head down or changing the patient's position to the supine or right lateral decubitus allows preferential right sided filling.
(g) To visualize the parts of common bile duct behind the duodenoscope, appropriate movements of the duodenoscope might be needed, such as advancing the scope in the longitudinal direction or withdrawing the duodenoscope into the stomach.
(h) Avoid injecting too much contrast from below the cystic duct insertion, as the gallbladder might be full.
(i) Use a balloon sweeping cholangiogram to evaluate distal strictures.
(j) Use the balloon occlusion method after stone clearance to obtain a good cholangiogram.

REFERENCES

[1] INTERNATIONAL COMMISSION ON RADIOLOGICAL PROTECTION, Avoidance of Radiation Injuries from Medical Interventional Procedures, Publication 85, Pergamon Press, Oxford and New York (2000).

[2] NATIONAL COUNCIL ON RADIATION PROTECTION MEASUREMENTS, NCRP Report No. 168: Radiation Dose Management for Fluoroscopically-Guided Interventional Medical Prcedures, NCRP, Bethesda (2010).

[3] INTERNATIONAL ATOMIC ENERGY AGENCY, Establishing and Improving Interventional Radiology, IAEA Human Health Series No. 42, IAEA, Vienna (2021).

[4] HELMBERGER, T., The evolution of interventional oncology in the 21st century, Br. J. Radiol. **93** (2020) 20200112,
https://doi:10.1259/bjr.20200112

[5] INTERNATIONAL COMMISSION ON RADIOLOGICAL PROTECTION, Radiological Protection in Cardiology, Publication 120, Pergamon Press, Oxford and New York (2013).

[6] PEPINE, C.J., et al., ACC/AHA guidelines for cardiac catheterization and cardiac catheterization laboratories. American College of Cardiology/American Heart Association Ad Hoc Task Force on Cardiac Catheterization, Circulation **84** (1991) 2213–2247,
https://doi.org/10.1161/01.CIR.84.5.2213

[7] HENDRY, C., RASHID, R., Catheter Laboratory Design, Staffing and Training, Primary Angioplasty: A Practical Guide (2018),
https://doi:10.1007/978-981-13-1114-7_6

[8] INTERNATIONAL ELECTROTECHNICAL COMMISSION, Medical Electrical Equipment-Part 2-54: Particular Requirements for the Basic Safety and Essential Performance of X-ray Equipment for Radiography and Radioscopy, IEC 60601-2-54, IEC, Geneva (2009).

[9] INTERNATIONAL ELECTROTECHNICAL COMMISSION, Medical Electrical Equipment–Part 2-43: Particular requirements for the basic safety and essential performance of X-ray equipment for interventional procedures, IEC 60601-2-43, IEC, Geneva (2010).

[10] RUIJTERS, D., HOMAN, R., MIELEKAMP, P., VAN DE HAAR, P., BABIC, D., Validation of 3-D multimodality roadmapping in interventional neuroradiology, Phys. Med. Biol. **56** (2011) 5335–5354,
https://doi:10.1088/0031-9155/56/16/017

[11] BUSHBERG, J.T., SEIBERT, J.A., LEIDHOLDT, E.M., BOONE, J.M., The Essential Physics of Medical Imaging, Wolters Kluwer Health, Philadelphia (2021).

[12] BALTER, S., Fluoroscopic technology from 1895 to 2019 drivers: Physics and physiology, Med. Phys. Int. **7** (2019) 111–140.

[13] NICKOLOFF, E.L., AAPM/RSNA physics tutorial for residents: physics of flat-panel fluoroscopy systems: survey of modern fluoroscopy imaging: flat-panel detectors versus image intensifiers and more, Radiographics **31** (2011) 591–602, https://doi:10.1148/rg.312105185

[14] BEVINS, N.B., et al., Practical application of AAPM Report 270 in display quality assurance: A report of Task Group 270, Med. Phys. **47** (2020) e920–e928, https://doi:10.1002/mp.14227

[15] SAMEI, E., et al., Assessment of display performance for medical imaging systems: executive summary of AAPM TG18 report, Med. Phys. **32** (2005) 1205–1225, https://doi:10.1118/1.1861159

[16] INTERNATIONAL COMMITTEE FOR DISPLAY METROLOGY, Information Display Measurements Standard (IDMS), version 1.1a, ICDM, (2022).

[17] DOWSETT, D., KENNY, P.A., JOHNSTON, R.E., The Physics of Diagnostic Imaging, CRC Press, Boca Raton, FL (2006), https://doi.org/10.1201/b13462

[18] BADAWY, M., CLARK, T., CARRION, D., DEB, P., FAROUQUE, O., Radiation dose optimization in interventional cardiology: A teaching hospital experience, Cardiol. Res. Pract. (2018), https://doi:10.1155/2018/6912841

[19] HIRSHFELD, J.W., et al, ACCF/AHA/HRS/SCAI clinical competence statement on physician knowledge to optimize patient safety and image quality in fluoroscopically guided invasive cardiovascular procedures: A report of the American College of Cardiology Foundation/American Heart Association/American College of Physicians Task Force on Clinical Competence and Training, Circulation **111(4)** (2005) 511–532, https://doi:10.1161/01.CIR.0000157946.29224.5D

[20] RUSSO, P., Handbook of X-ray Imaging: Physics and Technology, CRC Press, Boca Raton, FL (2018), https://doi.org/10.1201/9781351228251

[21] PERRY, B.C., MONROE, E.J., MCKAY, T., KANAL, K.M., SHIVARAM, G., Pediatric percutaneous osteoid osteoma ablation: Cone-beam CT with fluoroscopic overlay versus conventional CT guidance, CardioVasc. Interventional Radiol. **40** (2017) 1593–1599, https://doi:10.1007/s00270-017-1685-2

[22] DANCE, D.R., CHRISTOFIDES, S., MAIDMENT, A.D.A, McLEAN, I.D., NG, K.H., Diagnostic Radiology Physics: A Handbook for Teachers and Students, IAEA, Vienna (2014).

[23] INTERNATIONAL COMMISSION ON RADIATION UNITS AND MEASUREMENTS, Fundamental Quantities and Units for Ionizing Radiation, ICRU report 85, ICRU, Oxford University Press, Bethesda, MD (2011).

[24] INTERNATIONAL COMMISSION ON RADIATION UNITS AND MEASUREMENTS, Patient Dosimetry for X Rays Used in Medical Imaging, ICRU report 74, ICRU, Oxford University Press, Bethesda, MD (2005).

[25] AMERICAN ASSOCIATION OF PHYSICISTS IN MEDICINE, Estimating Patient Organ Dose with Computed Tomography: A Review of Present Methodology and Required DICOM Information. A Joint Report of AAPM Task Group 246 and the EFOMP, AAPM, Alexandria, VA (2019).

[26] OMAR, A., BUJILA, R., FRANSSON, A., ANDREO, P., POLUDNIOWSKI, G., A framework for organ dose estimation in x-ray angiography and interventional radiology based on dose-related data in DICOM structured reports, Phys. Med. Biol. **61** (2016) 3063,
https://doi:10.1088/0031-9155/61/8/3063

[27] HELLSTRÖM, M., GRANBERG, C., LUNDMAN, J., ANDERSSON, J., Estimating patient peak skin dose with fluoroscopic procedures, Eur. J. Med. Phys. **52** (2018) 105,
https://doi:10.1016/j.ejmp.2018.06.35

[28] SECHOPOULOS, I., TRIANNI, A., PECK, D., The DICOM radiation dose structured report: What it is and what it is not, J. Am. Coll. Radiol. **12** (2015) 712–713,
https://doi:10.1016/j.jacr.2015.04.002

[29] TSALAFOUTAS, I.A., KHARITA, M.H., AL-NAEMI, H., KALRA, M.K., Radiation dose monitoring in computed tomography: Status, options and limitations, Eur. J. Med. Phys. **79** (2020) 1–15,
https://doi:10.1016/j.ejmp.2020.08.020

[30] TSALAFOUTAS, I.A., et al, Technical specifications of dose management systems: An International Atomic Energy Agency Survey, J. Appl. Clin. Med. Phys. **25** 1 (2023), e14219,
https://doi:0.1002/acm2.14219

[31] WORLD HEALTH ORGANIZATION, Quality Assurance in Diagnostic Radiology, WHO, Geneva (1982).

[32] INTERNATIONAL ELECTROTECHNICAL COMMISSION, Evaluation and Routine Testing in Medical Imaging Departments Part 3-8: Acceptance and Constancy Tests — Imaging Performance of X-ray Equipment for Radiography and Radioscopy, IEC 61223-3-8, IEC, Geneva (2022).

[33] INTERNATIONAL ATOMIC ENERGY AGENCY, WORLD HEALTH ORGANIZATION, Bonn Call for Action: 10 Actions to Improve Radiation Protection in Medicine in the Next Decade, IAEA, Vienna (2014).

[34] EUROPEAN COMMISSION, FOOD AND AGRICULTURE ORGANIZATION OF THE UNITED NATIONS, INTERNATIONAL ATOMIC ENERGY AGENCY, INTERNATIONAL LABOUR ORGANIZATION, OECD NUCLEAR ENERGY AGENCY, PAN AMERICAN HEALTH ORGANIZATION, UNITED NATIONS ENVIRONMENT PROGRAMME, WORLD HEALTH ORGANIZAION, Radiation Protection and Safety of Radiation Sources: International Basic Safety Standards, IAEA Safety Series No. GSR Part 3, IAEA, Vienna (2014).

[35] INTERNATIONAL ATOMIC ENERGY AGENCY, Radiation Protection and Safety in Medical Uses of Ionizing Radiation, IAEA Safety Standards Series No. SSG-46, IAEA, Vienna (2018).

[36] INTERNATIONAL ATOMIC ENERGY AGENCY, Governmental, Legal and Regulatory Framework for Safety, IAEA Safety Standards Series No. GSR Part 1 (Rev. 1), IAEA, Vienna (2016).

[37] INTERNATIONAL ATOMIC ENERGY AGENCY, Roles and Responsibilities, and Education and Training Requirements for Clinically Qualified Medical Physicists, IAEA Human Health Series No. 25, IAEA, Vienna (2013).

[38] TSAPAKI, V., Radiation dose optimization in diagnostic and interventional radiology: Current issues and future perspectives, Eur. J. Med. Phys. **79** (2020) 16–21, https://doi:10.1016/j.ejmp.2020.09.015

[39] MEDICAL DEVICES AGENCY, Medical Device & Equipment Management for Hospital and Community-based Organisations, MDA Device Bulletin DB9801, MDA, London (1998).

[40] EUROPEAN COMMISSION, Radiation Protection No.162: Criteria for Acceptability of Medical Radiological Equipment used in Diagnostic Radiology, Nuclear Medicine and Radiotherapy, EC, Brussels (2012).

[41] AMERICAN ASSOCIATION OF PHYSICISTS IN MEDICINE, Quality Control in Diagnostic Radiology, Report No. 74, AAPM, Madison (2002).

[42] INSTITUTE OF PHYSICS AND ENGINEERING IN MEDICINE, Recommended Standards for the Routine Performance Testing of Diagnostic X-ray Imaging Systems, IPEM Report 91, IPEM, York, UK (2005).

[43] INSTITUTE OF PHYSICS AND ENGINEERING IN MEDICINE, Recommended Standards for the Routine Performance Testing of Diagnostic X Ray Imaging Systems, IPEM Report 77, IPEM, York, UK (1997).

[44] INTERNATIONAL ATOMIC ENERGY AGENCY, Handbook of Basic Quality Control Tests for Diagnostic Radiology, IAEA Human Health Series No. 47, IAEA, Vienna (2023), https://doi.org/10.61092/iaea.rhjo-8d0u

[45] INTERNATIONAL ELECTROTECHNICAL COMMISSION, IEC 62563-1:2009: Medical Electrical Equipment — Medical Image Display Systems — Part 1: Evaluation Methods, IEC, Geneva (2016).

[46] BEVINS, N.B., et al., Practical application of AAPM Report 270 in display quality assurance: A report of Task Group 270, Med. Phys. **47** (2020) e920–e928, https://doi:10.1002/mp.14227

[47] AMERICAN ASSOCIATION OF PHYSICISTS IN MEDICINE, Assessment of Display Performance for Medical Imaging Systems, AAPM Task Group Report 18, AAPM, Madison, WI (2005).

[48] MATHIEU, K.B., KAPPADATH, S.C., WHITE, R.A., ATKINSON, E.N., CODY, D.D., An empirical model of diagnostic x-ray attenuation under narrow-beam geometry, Med. Phys. **38** (2011) 4546–4555, https://doi:10.1118/1.3592933

[49] WUNDERLE, K.A., GODLEY, A.R., SHEN, Z.L., RAKOWSKI, J.T., DONG, F.F., Percent depth doses and X-ray beam characterizations of a fluoroscopic system incorporating copper filtration, Med. Phys. **44** (2017) 1275–1286, https://doi:10.1002/mp.12109

[50] INTERNATIONAL ELECTROTECHNICAL COMMISSION, Medical Electrical Equipment — Dose Area Product Meters, IEC 60580, IEC, Geneva (2019).

[51] NATIONAL COUNCIL ON RADIATION PROTECTION AND MEASURENENTS, Structural Shielding Design for Medical X ray Imaging Facilities, NCRP Report No. 147, NCRP, Bethesda, MD (2004).

[52] LIN, P.P., et al., AAPM Task Group Report 272: Comprehensive acceptance testing and evaluation of fluoroscopy imaging systems, Med. Phys. **49** (2022) e1–e49, https://doi:10.1002/mp.15429

[53] DICOM STANDARDS COMMITTEE, DICOM PS3.14 2023e – Grayscale Standard Display Function, NEMA, Arlington, VA (2023), https://dicom.nema.org/medical/dicom/current/output/chtml/part14/ps3.14.html

[54] PARUCCINI, N., et al., A single phantom, a single statistical method for low-contrast detectability assessment, Eur. J. Med. Phys. **91** (2021) 28–42, https://doi:10.1016/j.ejmp.2021.10.007

[55] ELGSTRÖM, H., TESSELAAR, E., SANDBORG, M., Signal-to-noise ratio rate measurement in fluoroscopy for quality control and teaching good radiological imaging technique, Radiat. Prot. Dosim. **195** (2021) 407–415, https://doi:10.1093/rpd/ncaa222

[56] TAPIOVAARA, M., Objective measurement of image quality in fluoroscopic X-ray equipment: FluoroQuality, Radiation and Nuclear Safety Authority, Vantaa, Finland (2003).

[57] TAPIOVAARA, M.J., SANDBORG, M., How should low-contrast detail detectability be measured in fluoroscopy? Med. Phys. **31** (2004) 2564–2576, https://doi:10.1093/rpd/ncaa222

[58] DAVIES, A.G., COWEN, A.R., KENGYELICS, S.M., BURY, R.F., BRUIJNS, T.J., Threshold contrast detail detectability measurement of the fluoroscopic image quality of a dynamic solid-state digital x-ray image detector, Med. Phys. **28** (2001) 11–15, https://doi:10.1118/1.1312814

[59] EVANS, D., MacKENZIE, A., LAWINSKI, C., SMITH, D., Threshold contrast detail detectability curves for fluoroscopy and digital acquisition using modern image intensifier systems, Br. J. Radiol. **77** (2004) 751–758, https://doi:10.1259/bjr/16707499

[60] GALLACHER, D., MacKENZIE, A., BATCHELOR, S., LYNCH, J., SAUNDERS, J., Use of a quality index in threshold contrast detail detection measurements in television fluoroscopy, Br. J. Radiol. **76** (2003) 464–472, https://doi:10.1259/bjr/82965103

[61] INTERNATIONAL ELECTROTECHNICAL COMMISSION, IEC 61910-1:2014: Medical Electrical Equipment — Radiation Dose Documentation — Part 1: Radiation Dose Structured Reports for Radiography and Radioscopy, IEC, Geneva (2014).

[62] EUROPEAN FEDERATION OF ORGANIZATIONS FOR MEDICAL PHYSICS, EUROPEAN RADIATION ONCOLOGY, INTERNATIONAL ATOMIC ENERGY AGENCY, Quality control in cone-beam computed tomography (CBCT) EFOMP-ESTRO-IAEA protocol, EFOMP, Utrecht, Netherlands (2019), https://doi:10.19285/CBCTEFOMP.V1.0.2017.06

[63] FERRARI, P., et al., EURADOS Working Group 12 studies in interventional radiology for medical staff dosimetry, Nuovo Cimento C **41**(6) (2018), https://doi:10.1393/ncc/i2018-18217-2

[64] AMERICAN ASSOCIATION OF PHYSICISTS IN MEDICINE, 3D C-arms with Volumetric Imaging Capability, AAPM Task Group Report 238, AAPM, Madison, WI (2023).

[65] INTERNATIONAL COMMISSION ON RADIOLOGICAL PROTECTION, The 2007 Recommendations of the International Commission on Radiological Protection. Publication 103, Pergamon Press, Oxford and New York (2008).

[66] INTERNATIONAL COMMISSION ON RADIOLOGICAL PROTECTION, Radiological Protection in Fluoroscopically Guided Procedures Performed Outside the Imaging Department. Publication 117, Pergamon Press, Oxford and New York (2010).

[67] INTERNATIONAL COMMISSION ON RADIOLOGICAL PROTECTION, Education and Training in Radiological Protection for Diagnostic and Interventional Procedures, Publication 113, Pergamon Press, Oxford and New York (2009).

[68] INTERNATIONAL COMMISSION ON RADIOLOGICAL PROTECTION, Radiological Protection in Paediatric Diagnostic and Interventional Radiology, Publication 121, Pergamon Press, Oxford and New York (2013).

[69] INTERNATIONAL COMMISSION ON RADIOLOGICAL PROTECTION, ICRP Statement on Tissue Reactions and Early and Late Effects of Radiation in Normal Tissues and Organs — Threshold Doses for Tissue Reactions in a Radiation Protection Context, Pergamon Press, Oxford and New York (2012).

[70] GINGOLD, E.L., The medical physicist's role in radiation optimization, J. Am. Coll. Radiol. **14** (2017) 1335–1336, https://doi:10.1016/j.jacr.2017.05.018

[71] SENSAKOVIC, W.F., WARDEN, D.R., BANCROFT, L.W., The link between radiation optimization and quality, J. Am. Coll. Radiol. **14** (2017) 850–851, https://doi:0.1016/j.jacr.2017.03.020

[72] INTERNATIONAL ATOMIC ENERGY AGENCY, Quality Assurance Programme for Screen Film Mammography, IAEA Human Health Series No. 2, IAEA, Vienna (2009).

[73] INTERNATIONAL ATOMIC ENERGY AGENCY, Dosimetry in Diagnostic Radiology: An International Code of Practice, Technical Reports Series No. 457, IAEA, Vienna (2007).

[74] INTERNATIONAL ATOMIC ENERGY AGENCY, Quality Assurance Programme for Digital Mammography, IAEA Human Health Series No. 17, IAEA, Vienna (2011).

[75] EUROPEAN COMMISSION, European Guidelines for Quality Assurance in Breast Cancer Screening and Diagnosis, 4th edn., EU Publications Office, Brussels (2013), https://data.europa.eu/doi/10.2772/13196

[76] AMERICAN ASSOCIATION OF PHYSICISTS IN MEDICINE, Acceptance Testing and Quality Control of Photostimulable Phosphor Imaging Systems, AAPM Report No. 93, AAPM, Madison, WI (2006).

[77] INTERNATIONAL COMMISSION ON RADIOLOGICAL PROTECTION, Diagnostic reference levels in medical imaging, Publication 135, Pergamon Press, Oxford and New York (2017).

[78] SCHEGERER, A.A., et al., Radiation dose and diagnostic reference levels for four interventional radiology procedures: results of the prospective European multicenter survey EUCLID, Eur. Radiol. **31** (2021) 9346–9360, https://doi:10.1007/s00330-021-08029-y

[79] DAMILAKIS, J., et al., How to establish and use local diagnostic reference levels: An ESR EuroSafe Imaging expert statement, Insights into Imaging **14** (2023) 27, https://doi:10.1186/s13244-023-01369-x

[80] EUROPEAN COMMISSION, Radiation Protection No, 195: European study on clinical diagnostic reference levels for X-ray medical imaging — EUCLID, Publications Office of the European Union, Luxembourg (2021), https://data.europa.eu/doi/10.2833/452154

[81] TSALAFOUTAS, I., TSAPAKI, V., TRIANTOPOULOU, I., Evaluation of image quality and patient exposure in fluoroscopy using a phantom: Is there any clinical relevance?, Eur. J. Radiol. **138** (2021) 109607, https://doi:10.1016/j.ejrad.2021.109607

[82] MALCHAIR, F., D 9.141-Standards for digital dose reporting, European Joint Programme for the Integration of Radiation Protection Research (2018), https://doi:10.3030/662287

[83] STECKER, M.S., et al., Guidelines for patient radiation dose management, J. Vasc. Interv. Radiol. **20** (2009) S263–S273, https://doi:10.1016/j.jvir.2009.04.037

[84] GEIJER, H., PERSLIDEN, J., Radiation exposure and patient experience during percutaneous coronary intervention using radial and femoral artery access, Eur. Radiol. **14** (2004) 1674–1680, https://doi:10.1007/s00330-004-2322-4

[85] TARIGHATNIA, A., ALIAN, A.H.M., GHOJAZADEH, M., FARAJOLLAHI, A.R., Comparison of the patient radiation exposure during coronary angiography and angioplasty procedures using trans-radial and trans-femoral access, J. Cardiovasc. Thorac. Res. **8**(2) (2016) 77–82, https://doi:10.15171/jcvtr.2016.15

[86] MANICARDI, M., et al., Anthropometric parameters and radiation doses during percutaneous coronary procedures, Eur. J. Med. Phys. **100** (2022) 164–175, https://doi:10.1016/j.ejmp.2022.06.013

[87] TSAPAKI, V., et al., Management of difficult bile duct stones and indeterminate bile duct structures: reduced ERCP radiation exposure with adjunct use of digital single-operator cholangioscopy, Eur. J. Med. Phys. **64** (2019) 69–73, https://doi:10.1016/j.ejmp.2019.06.002

[88] HAYASHI, S., et al., Diagnostic reference levels for fluoroscopy-guided gastrointestinal procedures in Japan from the REX-GI study: A nationwide multicentre prospective observational study, Lancet Reg. Health West. Pac. **20** (2022) 100376, https://doi:10.1016/j.lanwpc.2021.100376

[89] TSAPAKI, V., et al., Preliminary diagnostic reference levels for endoscopic retrograde cholangio-pancreatography in Greece, Eur. J. Med. Phys. **32** (2016) 607–611, https://doi:10.1016/j.ejmp.2016.03.018

[90] SAUKKO, E., HENNER, A., NIEMINEN, M.T., AHONEN, S.M., The establishment of local diagnostic reference levels in endoscopic retrograde cholangiopancreatography: A practical tool for the optimisation and for quality assurance management, Radiat. Prot. Dosimetry **173** (2017) 338–344, https://doi:10.1093/rpd/ncw018

[91] DUMONCEAU, J.M., et al., Radiation protection in digestive endoscopy: European Society of Digestive Endoscopy (ESGE) guideline, Endoscopy **44** (2012) 408–421, https://doi:10.1055/s-0031-1291791

[92] JORGENSEN, J.E., RUBENSTEIN, J.H., GOODSITT, M.M., ELTA, G.H., Radiation doses to ERCP patients are significantly lower with experienced endoscopists, Gastrointestinal Endoscopy **72** (2010) 58–65, https://doi:10.1016/j.gie.2009.12.060

[93] LIAO, C., et al., Radiation exposure to patients during ERCP is significantly higher with low-volume endoscopists, Gastrointest. Endosc. **81** (2015) 391–398, https://doi:10.1016/j.gie.2014.08.001

[94] KATSINELOS, P., et al., A prospective analysis of factors influencing fluoroscopy time during therapeutic ERCP, Ann. Gastroenterol. **25** (2012) 338.

[95] OH, C.H., SON, B.K., Minimizing radiation exposure in endoscopic retrograde cholangiopancreatography: A review for medical personnel, Korean J. Intern. Med. **37** (2022) 1111, https://doi:10.3904/kjim.2022.093

[96] BOWSHER, W., BLOTT, P., WHITFIELD, H., Radiation protection in percutaneous renal surgery, British J. Urol. **69** (1992) 231–233, https://doi:10.1111/j.1464-410x.1992.tb15518.x

[97] IMAI, S., et al., Occupational eye lens dose in endoscopic retrograde cholangiopancreatography using a dedicated eye lens dosimeter, J. Radiol. Prot. **41(3)** (2021), https://doi:0.1088/1361-6498/ac091f

[98] DRAGANOV, P.V., FORSMARK, C.E., Prospective evaluation of adverse reactions to iodine-containing contrast media after ERCP, Gastrointest. Endosc. **68**(6) (2008) 1098–1101, https://doi:10.1016/j.gie.2008.07.031

[99] GARDNER, T.B., BARON, T.H., Optimizing cholangiography when performing endoscopic retrograde cholangiopancreatography, Clin. Gastroenterol. Hepatol. **6**(7) (2008) 734–740, https://doi:10.1016/j.cgh.2008.04.015

ABBREVIATIONS

AAPM	American Association of Physicists in Medicine
AERC	automatic exposure rate control
BMI	body mass index
CAU	caudal
CBCT	cone beam computed tomography
CQMP	clinically qualified medical physicist
CRA	cranial
CRT	cathode ray tube
CT	computed tomography
DACS	dose archiving and communication systems
DICOM	digital imaging and communications in medicine
DRL	diagnostic reference level
DSA	digital subtraction angiography
EFOMP	European Federation of Organisations for Medical Physics
ERCP	endoscopic retrograde cholangiopancreatography
ESAK	entrance surface air kerma
ESTRO	European Society for Therapeutic Radiology and Oncology
FGIP	fluoroscopically guided interventional procedure
FOV	field of view
FPD	flat panel detector
fps	frames per second
GSDF	grey scale display function
HVL	half value layer
IEC	International Electrotechnical Commission
IPEM	Institute of Physics and Engineering in Medicine
kVp	peak kilovoltage
LAO	left-anterior-oblique
LCD	liquid crystal display
LIH	last image hold
MRT	medical radiation technologist
OLED	organic light-emitting diode
PACS	picture archiving and communication systems
PMMA	polymethyl methacrylate
pps	pulses per second
QA	quality assurance
QAP	quality assurance programme
QC	quality control
RAO	right-anterior-oblique

RDSR	radiation dose structured report
SID	source-to-image distance
SNR	signal to noise ratio
TCDD	threshold contrast detail detectability
UPS	uninterrupted power supply

CONTRIBUTORS TO DRAFTING AND REVIEW

Abdullah, B.	PJ Centre for Advanced Surgery and Oncology, Malaysia
Alvarez, W.	Fundación Valle del Lili Hospital, Colombia
Balter, S.	Columbia University Medical Center, United States of America
Brady, Z.	Alfred Health, Australia
Carrara, M.	International Atomic Energy Agency
de Vos, H.J.	Netcare Hospitals, South Africa
Gkolfakis, K.	Konstantopoulio General Hospital, Greece
Khoury, H.	Universidade Federal de Pernambuco, Brazil
Maccia, C.	Centre d'assurance de qualité des applications technologiques, France
Malchair, F.	Zephyra, Belgium
Mikhail, M.	International Atomic Energy Agency
Paraskeva, K.	Konstantopoulio General Hospital, Greece
Rizk, C.	International Atomic Energy Agency
Schueler, B.	Mayo Clinic, United States of America
Trianni, A.	Santa Chiara Hospital, Italy
Tsapaki, V.	International Atomic Energy Agency
Ubeda, C.	Universidad de Tarapacá, Chile
Wunderle, K.	Ohio State University Wexner Medical Center, United States of America
Yeong Chaihong	Taylor's University, Malaysia

Consultants Meetings

Vienna, Austria: 9–13 May 2022, 17–21 April 2023

IAEA
International Atomic Energy Agency

CONTACT IAEA PUBLISHING

Feedback on IAEA publications may be given via the on-line form available at:
www.iaea.org/publications/feedback

This form may also be used to report safety issues or environmental queries concerning IAEA publications.

Alternatively, contact IAEA Publishing:

Publishing Section
International Atomic Energy Agency
Vienna International Centre, PO Box 100, 1400 Vienna, Austria
Telephone: +43 1 2600 22529 or 22530
Email: sales.publications@iaea.org
www.iaea.org/publications

Priced and unpriced IAEA publications may be ordered directly from the IAEA.

ORDERING LOCALLY

Priced IAEA publications may be purchased from regional distributors and from major local booksellers.

MIX
Papier | Fördert gute Waldnutzung
FSC® C083411

Zeitfracht Medien GmbH
Ferdinand-Jühlke-Straße 7
99095 Erfurt, Deutschland
produktsicherheit@kolibri360.de